Creative City as an Urban Development Strategy

"This compendium is an invaluable read for urban geographers/planners. The Malaysian government has formulated a set of strategies to enhance urban development and the role of cities. This book contributes profoundly to this objective. It rises above previous work on competitive cities in Malaysia in terms of detail and richness of local knowledge."

—Morshidi Sirat, *Professor of Urban Geography, Universiti Sains Malaysia*

"Ipoh is synonymous with an ideal retirement city and old town for many people. In fact, its rich history and cultural heritage lay good potential and endless possibilities for urban regeneration. This book is definitely providing a timely analysis and insightful perspective for us to understand more about this former tin mining city, and on efforts to develop it towards a creative city."

—Tan Kar Hing, *former Perak state's executive member for Tourism, Arts and Culture*

Suet Leng Khoo • Nicole Shu Fun Chang

Creative City as an Urban Development Strategy

The Case of Selected Malaysian Cities

palgrave
macmillan

Suet Leng Khoo
School of Social Sciences
Universiti Sains Malaysia
Penang, Malaysia

Nicole Shu Fun Chang
Equator College
Penang, Malaysia

ISBN 978-981-16-1290-9 ISBN 978-981-16-1291-6 (eBook)
https://doi.org/10.1007/978-981-16-1291-6

This Palgrave Macmillan imprint is published by the registered company Springer Nature Singapore Pte Ltd.
The registered company address is: 152 Beach Road, #21-01/04 Gateway East, Singapore 189721, Singapore

Preface

The idea of writing this book was conceptualised in 2018 after I have indulged in a myriad of research projects related to creative city. As I delved deeper into the creative city discourse, there were more and more 'insights' garnered as well as 'enquiries and ambiguities' that surfaced in terms of fathoming the concept from practical, theoretical and policy viewpoints. This made me realise that investigating these enquiries cannot be merely captured in a journal manuscript but would require more audacious endeavours like writing a book. The aspiration to pen this book gradually morphed into reality after a casual conversation with my co-author, Nicole Chang Shu Fun, a budding scholar in the field who enthusiastically accepted the invitation to embark on this book project with me.

After undertaking extensive research in our home city George Town (capital of Penang) as testbed of the concept, the time is also opportune for us to transcend national boundaries and test the concept for other Malaysian cities. In recent years, the accelerated fashion in which creative city development has reached and anchored on Malaysian shores has further instigated us to explore this topic. The adoption and adaptation of this concept within Malaysian soils will in one way or another shape Malaysia's urban development, policy discourses and thus warrant immediate attention and concerted actions from all stakeholders. These exigencies have also prompted us to systematically test out this urban development strategy for other Malaysian cities besides George Town which is a UNESCO World Heritage Site.

Additionally, the motivation to pen this book is also provoked by the paradoxical convictions of the concept itself where its acceptance is contentious and ambivalent. It is a concept that is both heralded and yet criticised. Globally, urban managers and policy-makers tend to advocate and promote creative city development as the way forward for their cities. Contrastively, scholars in the academic realm are rather sceptical about the concept and often label it as elitist, divisionary and exacerbating urban exclusion. It is against this dialectical backdrop that the creative city concept is being hypothesised and explored for four selected Malaysian cities, namely, Kuala Lumpur, George Town, Ipoh and Johor Bahru.

From 2018 to 2020, primary data collection (i.e. interviews, focus group discussions, fieldwork observations, etc.) was undertaken in these four cities and the data set was supplemented by extensive secondary data collection (i.e. technical reports, development blueprints, journals, etc.). By triangulating primary and secondary data, we managed to chronicle and depict existing creative city initiatives and pursuits embraced by each of the Malaysian cities, analysed each city's viability, identified their challenges and also examined the roles of each city's development actors in propounding and embracing this strategy. In testing the applicability and viability of the creative city concept, we also delved deep into deciphering, deconstructing and reconceptualising the concept to fit local Malaysian nuances and idiosyncrasies. With Malaysia's macro shift and strategic direction re-orientating towards the creative economy, the writing of this book is apt and timely to dissect and comprehend the way creative and cultural industries are fused within Malaysia's urban development policy for growth and development at the local, national and regional level. It is hoped that the narratives, recommendations and insights generated from this book will be useful for Malaysian policy-makers, urban managers, private entrepreneurs, scholars and civil society alike as we strive to model a Malaysian creative city that is inclusive, resilient, proactive and sustainable so that no one will be left behind.

Penang, Malaysia
December 2020

Suet Leng Khoo

ACKNOWLEDGEMENTS

The writing of this book has been both enriching and challenging. The fieldwork trips were particularly insightful and rewarding where we visited new places and met new faces along the way. Data collection in George Town and outstation fieldwork trips to Kuala Lumpur, Ipoh and Johor Bahru have indeed broadened our horizons and widened our circle of networks and contacts. Although the two authors are responsible for the text and content of the book, this book could not have been written without the kind assistance and cooperation of others. Their cooperation and contributions are invaluable. Thus, we would like to extend our heartfelt thanks to all parties, organisations, institutions and individuals who have contributed in one way or another towards the successful production and publication of this book. We are also extremely grateful to all key stakeholders and informants who participated in our interviews and focus group discussions in the course of writing this book. We also take this opportunity to thank our family members for their patience, support and motivation when we were busy working on this book during the past two years.

Contents

About the Authors

Suet Leng Khoo (PhD, MMIP) is an Associate Professor in the Department of Development Planning and Management, School of Social Sciences, Universiti Sains Malaysia. Dr Khoo has vast national and international experiences on creative city research. In 2015, she collaborated with Think City Sdn Bhd (urban regeneration think-tank) to prepare a baseline study of Penang's creative and cultural sector. From 2017 to 2019, she partnered with the University of Greenwich, United Kingdom, through the Newton-Ungku Omar Fund to study George Town World Heritage Site's creative and cultural value chain. In 2018, Dr Khoo received an international grant from the Sumitomo Foundation (Japan) where she visited the Creative City of Kanazawa for data collection. Dr Khoo was Consultative Panel member for George Town World Heritage Incorporated from 2011 to 2015 and Technical Advisory Panel member for Think City Sdn Bhd since 2015. Dr Khoo is a registered Town Planner with the Malaysian Board of Town Planners and a corporate member in the Malaysian Institute of Planners.

Nicole Shu Fun Chang (PhD) completed her PhD in 2019. Dr Chang's research investigated the potential of visual arts as one of the core creative industries and the role of visual arts in public domain. From 2003 to 2006, she was actively involved in academic research as a research officer in Universiti Sains Malaysia in the fields of property management and urban conservation. Currently, Dr Chang is a freelance consultant where she does project-based writing and research about local place-based initiatives in creative place-making. She writes and undertakes research for both private-commissioned projects and government-funded projects in Penang. She frequently contributes articles to *Penang Monthly* (a publication by Penang Institute – Penang state government's think-tank) and Penang Art District's blog (a blog managed by Penang Art District – Penang state government's planned hub of contemporary arts and culture). Dr Chang also lectures on a part-time basis in Equator College (an art, media and design college in Penang).

Abbreviations

ASEAN	Association of Southeast Asian Nations
ASEF	Asia-Europe Foundation
CAT	Creative Animation Triggers
CBD	Central Business District
CCI	Creative and cultural industries
CDP	Comprehensive Development Plan
CENDANA	Cultural Economy Development Agency
CI	Creative Industries
CMC	Creative Multimedia Content
CMCO	Conditional Movement Control Order
Covid-19	Corona virus disease
DCMS	Department of Culture, Media and Sport (UK government)
DNA	Deoxyribonucleic acid
DoSM	Department of Statistics, Malaysia
EIC	East-Indian Company
ERC-CI	Economic Review Committee – Services Subcommittee Workgroup on Creative Industries
ERL	Express Rail Link Sdn. Bhd.
EXCO	Executive Council
F & B	Food & Beverages
FGD	Focus group discussion
GATS	General Agreement on Trade in Services
GDP	Gross Domestic Product
GII	Global Innovation Index
GLAM	Galleries, Libraries, Archives and Museums
GNI	Gross National Income
GNP	Gross National Product

GT	George Town
GTF	George Town Festival
GTWHI	George Town World Heritage Incorporated
GTWHS	George Town World Heritage Site
Habitat	United Nations Conference on Housing and Sustainable Urban Development
HSBC	Hongkong and Shanghai Banking Corporation
HUL	Historic Urban Landscape
ICC	Ipoh City Council
ICT	Information and communications technology
IIAF	Ipoh International Art Festival
IM	Iskandar Malaysia
IMS-GT	Indonesia-Malaysia-Singapore Growth Triangle
IRDA	Iskandar Regional Development Authority
ITHC	Ipoh Tin Heritage City
JB	Johor Bahru
JBCC	Johor Bahru City Council
JSPA	Johor Society for the Performing Arts
KL	Kuala Lumpur
KLCCD	Kuala Lumpur Creative and Cultural District
KLIA 1	Kuala Lumpur International Airport 1
KLIA 2	Kuala Lumpur International Airport 2
KRI	Khazanah Research Institute
KTM	Keretapi Tanah Melayu
LRT	Light Rail Transit
MATRADE	Malaysia External Trade Development Corporation
MCM	Ministry of Communications and Multimedia
MCMC	Malaysian Communications and Multimedia Commission
MCO	Movement Control Order
MDeC	Multimedia Development Corporation
MICC	Ministry of Information, Communications and Culture
MITA	Ministry of Information and the Arts
MOTAC	Ministry of Tourism, Arts and Culture, Malaysia
MSC	Multimedia Super Corridor
MSIC	Malaysia Standard Industrial Classification
MyPAA	My Performing Arts Agency
NCIP	National Creative Industry Policy
NCP	National Cultural Policy
NEP	New Economic Policy
NGO	Non-governmental organisation
NUP 2	National Urbanisation Policy 2
OECD	The Organisation for Economic Co-operation and Development

OUV	Outstanding Universal Value
PAD	Penang Art District
PARP	Public Art Review Panel
PDC	Penang Development Corporation
PETACE	Penang State Executive Council (EXCO) for Tourism and Creative Economy
PETACH	Penang State Executive Council (EXCO) for Tourism Development, Arts, Culture and Heritage
PGT	Penang Global Tourism
PORT	People of Remarkable Talents
RM	Malaysian ringgit
SAP	Special Area Plan
SDGs	Sustainable Development Goals
SEA	Southeast Asia
SUPERI	Sungai Perak Indah
TCPD	Town and Country Planning Department
TCSB	Think City Sdn Bhd
TFCC	Tom Fleming Creative Consultancy
UCCN	UNESCO Creative Cities Network
UK	United Kingdom
UNCTAD	United Nations Conference on Trade and Development
UNDP	United Nations Development Programme
UNESCO	United Nations Educational, Scientific and Cultural Organization
USD	US Dollar
WHS	World Heritage Site
WIPO	World Intellectual Property Organization

List of Figures

List of Tables

CHAPTER 1

Introduction

Abstract This introductory chapter will set the backdrop and discuss the motivation to write this book. It will begin by briefly illustrating the growing prominence of creative city discourse globally (to serve as foreground) before fleshing out the need to explore this strategy for the Malaysian context. This chapter will also provide a snapshot of Malaysia's overall development trajectory, the nation's strategic directions and the way they impact urban development policies thus exploring the notion of creative city as a viable strategy for Malaysian cities.

Keywords Creative cities • Malaysia • Urban development • Urban policy • Development trajectory

1.1 Nascency of Creative City Development in Malaysia

The concept of creative city has assumed centrality in urban policy-making and scholastic discourse in the last few decades due to structural economic shifts and global urban restructuring. Arguably, the original role of cities as sites of industrial manufacturing has been displaced and replaced with services, cultural and knowledge-based activities. As a result, cities world-wide are now restrategising to harness their cultural and creative endowments for local economic development (Boccella & Salerno, 2016) and to

S. L. Khoo, N. S. F. Chang, *Creative City as an Urban Development Strategy*, https://doi.org/10.1007/978-981-16-1291-6_1

enhance their competitive edge and comparative advantage. While much of the early literature originated from North America and Western Europe (Landry & Bianchini, 1998; Landry, 2005, 2008; Mercer, 2006; Smidt-Jensen, 2007), the notion of creative city transcends beyond Western political boundaries given that each city is, by default, culturally endowed and creative in its own way. Hence, what is observed is the way each urban area embodies and embraces the cultural and creative turn in its respective cities.

As the creative city hype moves across the globe, many urban managers especially in the Asian region (i.e. Korea, China, Indonesia, Singapore, Taiwan, Hong Kong) have endorsed it as a viable urban development strategy (UNCTAD, 2017; Hidayat and Asmara, 2017; Chaudhury & Lundbery, 2018; Karvelyte, 2018). This is despite critiques, predominantly from the academic realm, that opposed the concept as elitist, divisionary and exclusive (McCann, 2007; Pratt, 2011; Lee, 2011). Malaysian cities too are no exception with attempts to adopt this concept though in a fragmented fashion. This will be deliberated in subsequent chapters. Inadvertently, the processes and depictions of what define a creative city differ considerably across geographical and temporal contexts which are predominantly shaped by a city's or nation's politico-institutional framework and historical legacies.

In Malaysia, the role of cities and the importance of urban development are increasingly moving to the forefront of policy agenda given that the share of urban population in Malaysia is 74.3% (22.66 million) in 2015 and projected to increase to 79.6% (27.30 million) in 2025 (TCPD, 2016). Major states like Kuala Lumpur, Penang, Perak and Johor are rapidly urbanising and modernising with urbanisation rates of 100%, 96.2%, 81% and 79.8% respectively in 2019 (DoSM, 2020). The common trends in most, if not all, Malaysian cities are urban restructuring processes which are necessitated by changing demographics, alterations to the city's economic base shifting from industrial to services and reorientation towards the creative economy. While there are possibly upsides to transform Malaysian cities into creative cities given that these endeavours permeate into the development agendas of cities like George Town, Kuala Lumpur, Ipoh and Johor Bahru, the process is also fraught with dilemmas that revolve around rudimentary yet fundamental issues such as what constitutes a Malaysian creative city, and the way cultural and creative industries/economy are delineated and defined in the Malaysian context. While Malaysia's ethnic plurality and cultural diversity are unique and can be leveraged to showcase the nation's creative and cultural endowments, at the same instance, these are latent and contentious elements to be reckoned with when addressing the enquiry of 'what, whose and which cultural/creative endowments need to be prioritised, showcased and safeguarded in Malaysian cities for urban liveability and posterity. The case cities in Chaps. 5, 6, 7 and 8 will deliberate these in length.

Thus, this book is a novel attempt to investigate the viability of creative city strategies for selected Malaysian cities as well as to fathom the dynamics and nexus between the state, market and civil society in the process. Extending from the viewpoints of Western and Eastern scholars as well as policy-makers, this book will delve deep into the conceptualisation of Malaysian creative cities (in-the-making) by unravelling the contested and debatable meanings of culture, creativity, cultural industries and creative industries. The divergence or convergence that emerge will be juxtaposed against generically accepted notions pertaining to the foundations, people, networks, planning, markets and visions that should underscore a holistic and inclusive Malaysian creative city. More distinctively, this book is a pioneering work to contextualise and reconceptualise the creative city notion for Malaysia, which is a unique country shaped by multiplicities of ethnicities, backgrounds, creed and cultures.

1.2 Creative City as Viable Urban Development Strategy

Since time immemorial, the creative city concept has existed but under different guises. The concept is generally viewed from two distinctively different yet intertwined notions of 'culture' and 'creativity'. At this juncture, amongst the biggest contention is to first fathom the delineation and differentiation between the notions of 'cultural industries' and 'creative industries'. Scholars have argued that cultural industries predate creative industries and the former are defined according to a combination of five main criteria, namely, creativity, intellectual property, use value, production methods and symbolic meaning. In particular, symbolic meaning is the defining concept of culture and the economic value generated will reflect their cultural value. Thus, the cultural industries should not be blatantly and carelessly be subsumed under the wider creative industries agenda due to heightened interest in the 'knowledge economy' (Galloway & Dunlop, 2007: 19, 25). This contention will plausibly alter the meaning and making of a creative city. This will be further dissected in Chap. 2. However, the embeddedness of creativity in urban settings to spur more creativity and innovation cannot be dismissed either. Arguably, the notion of creative city dates back 2500 years ago when cities like classical Athens, renaissance Florence, enlightenment London and Fin de Siecle Vienna were the 'hotbeds of revolutionary creativity' (Andersson, 2011: 39). These cities functioned as magnets to attract migrants with creative prowess and to nurture creativity development among existing city dwellers. These were also sites to showcase discoveries from elsewhere and they doubled up as conducive locales to conceive new ideas together with other creative people. Andersson further argues that the formulaic blend of the right pool and interactions of creative human capital will consolidate overall creativity. This is further reinforced by an enabling environment or a 'creative milieu' that is large, deep, diverse and broad in cultural scope (Andersson, 2011: 51).

Extending the narrative from medieval times as espoused above, the creative city concept morphed and subsequently gathered attention through Peter Hall's seminal work 'Cities in Civilization: Culture, Innovation and Urban Order'; and also, Jane Jacobs' conviction of a city's diversity and its role as engine of economic growth. In her work 'Economy of Cities', Jacobs discerned cities as a co-existence and symbiosis of multiple levels/layers of clusters, intersected labour force, constant immigration of labour and also a diverse and rich milieu. Jacobs further highlighted that this 'rich milieu' will then affect and condition the birth of novel ideas, skills and knowledge that instil creativity and innovation in industries, firms and individuals (Jacobs, 1972). Later, more contemporary urban theorists like Charles Landry (2005, 2008) and Richard Florida (2002, 2004, 2007, 2008) have made popular the concept though much critiques have been levelled against Florida's elitist 'creative class' terminology. While the notion of creative city has gained credence worldwide, especially amongst urban managers and policy-makers in cities like Toronto, Liverpool, Melbourne, Seoul, Curitiba and Aswan[1] who embraced this strategy as part of their urban development strategy, a lesser level of acceptance derives from the academic realm where scholars are ambivalent with its adoption given the conceptual ambiguities, blurry delineation between culture and creativity (Galloway & Dunlop, 2007)

[1] Creative Cities Network. Source: https://en.unesco.org/creative-cities/creative-cities-map (accessed on 19 September 2019).

and the potential and new forms of divides created by advancing this agenda (Pratt, 2011).

On the Malaysian forefront, however, the development of creative cities and the emphasis on 'creativity' and 'innovation' are aligned to the nation's broader development vision (i.e. Vision 2020) in the quest to achieve a developed nation status by 2020. Arguably, the Malaysian process inclines towards viewing creative industries and city development as part of the knowledge-based economy agenda instead of advancing cultural creativity and cultural contents of the cultural industries. During the Prime Minister's Innovation Awards held in 2015, former Prime Minister Najib Razak highlighted that innovation and creativity are the two most crucial elements in Malaysia's development to achieve Vision 2020 and urged all relevant stakeholders to emphasise on innovation and creativity during brainstorming sessions.[2] It was again reiterated in the following year that Malaysia's new direction after Vision 2020 will focus on the need for innovation and creativity in order to survive in a highly competitive global world.[3]

1.3 The Growing Importance of Creative Economy in Malaysia

Before we delve into reconceptualising creative city for the Malaysian context, it is pivotal to briefly understand the nation's structural economic shift and overall development which underpin the process.

1.3.1 Malaysia's Overall Development Trajectory

Since independence, Malaysia has progressively morphed from a rural, traditional and relatively poor nation to a fairly prosperous country (Lafaye de Micheaux, 2017). Since 1994, Malaysia has progressed from a lower middle-income country to be an upper middle-income country as per the World Bank's categorisation based on per capita gross national income

[2] Innovation, creativity vital to achieve vision 2020, says Najib. Source: https://mymagic.my/news/innovation-creativity-vital-to-achieve-vision-2020-says-najib/ (accessed on 19 September 2019).

[3] PM: Innovation and creativity a necessity for Malaysia to survive post 2020. Source: https://www.nst.com.my/news/2015/12/117574/pm-innovation-and-creativity-necessity-malaysia-survive-post-2020 (accessed on 19 September 2019).

(GNI). Subsequently, a World Bank's report in 2017 predicted that Malaysia will reach the high-income threshold between 2020 and 2024 (World Bank, 2017).

Tracing Malaysia's economic growth trajectory since 1960, there were distinctive structural economic transformations that responded to the different economic conditions and demands of each progression. From 1960 to 1985, the commodity-based economy has slowly began to industrialise and evolved into an export-oriented manufacturing economy. Export-oriented industrialisation has been flourishing in full swing throughout the period of 1986 to 1998 (KRI, 2018). After major economic turbulence like the Asian and Global Financial Crises, the Malaysian economy restructured towards enhancing resilience and robustness of the industrial sector. Since 1998, Malaysia's economy started to deindustrialise and focused on diversifying the sectoral development until today. The transformation and growth of the nation were reflected from the decreasing average unemployment rate as well as the growth of national GDP and annual median household income throughout the different eras. KRI (2018) reported that Malaysia's real GDP has grown from RM69 billion in 1970 to RM1.1 trillion in 2016 while Malaysian households earned higher income throughout the years. According to KRI (2018), throughout the industrialisation phase, the real monthly median household income has rapidly rose from RM711 in 1970 to RM2,466 in 2002 and later reached RM4,538 in 2016.

Malaysia's economic transformation has direct impact on the change of employment involvement by sectors. According to KRI (2017), it was reported that the share of employment in agriculture sector has shrunk since 1970 while Malaysians working in industrial and services sectors began to increase. Until 2000, the industrial sector's employment share stopped growing whereas the services sector's employment share has been increasing until today to support Malaysia's diversified industrial development.

In order to transit from a middle-income to high-income economy, Malaysia must step beyond being a hub of production towards becoming a hub of creation (KRI, 2018). According to Aghion and Bircan (2017), the said economic transition involved the shift from investment-led growth that centred on capital accumulation and technology adoption to innovation-led growth that emphasises productivity enhancements and new value creation. This implies the crucial need of investment to develop knowledge and innovation towards building a creative society to drive Malaysia's future development on the global platform.

1.3.2 Creative Sectors as New Economic Catalyst in Malaysia

In the last decade, the creative sectors are earmarked as Malaysia's new economic catalyst. As reported in the Global Innovation Index (GII) 2016 (Cornell University, INSEAD and WIPO, 2016), Malaysia scored 43.4 and ranked at 35th for innovation. With this score, Malaysia was ranked at the second top among upper-middle income economies. The rankings unleashed Malaysia's potential to compete in the global innovative and creative market towards achieving high-income status. Under the specific pillars of innovation, Malaysia ranked 34th for human capital and research; 35th for knowledge and technology outputs; and 43rd for creative outputs (Cornell University, INSEAD and WIPO, 2016). From these measurements, the 'creative industries' label in Malaysia (that will underpin a creative city) skews heavily towards the knowledge economy model akin to Howkins' (2002) (cited in Galloway & Dunlop, 2007: 20) propositions that 'creative industry' would refer to industries where brain power is 'preponderant' and 'intellectual property' is the outcome.

The growing importance of creativity and innovation have motivated policy-makers to undertake research related to creative industries/creative cities and to develop Malaysia's creative economy. At the earlier stage, the Malaysian government initiated the Creative Industry Baseline Study 2008 for the Multimedia Development Corporation (MDeC). It studied the creative content industries which comprise of advertisement, animation, digital content and application, film and TV, games, music and radio. Based on the study, creative industries (focusing on creative content industries) have contributed RM9.4 billion (1.27%) to the national GDP, RM1.4 billion of export and have a 45,301 workforce (0.4%) in 2008 (Hasnul, 2011; Hisham, 2016). Since then, the Malaysian creative content industries have started to contribute more significantly to the national revenue. For instance, its contribution of 1.5% (RM11.6 billion) in 2010 (Lim, 2014) and 1.6% (RM16 billion) in 2014 (Digital News Asia, 2014) to Malaysia's GDP. However, Malaysia External Trade Development Corporation (MATRADE) (2017) reported that the said sector contributed 6.4% (RM9.55 billion) of the total GDP in 2015. In 2017, MATRADE reported that Malaysia has exported a total of RM6.8 billion worth of creative contents until January 2017. When comparing the earlier figures of contributions (in ringgit Malaysia) as recorded throughout the different years, data inconsistency is found. This somewhat implies the inconsistency of industrial categorisation in Malaysia, especially under the creative sector which consists of a wide scope of industries.

Apart from data inconsistency, the contribution of Malaysia's creative industries in comparison to other developed economies is still minimal. In 2012, MyCreative Ventures Sdn Bhd was launched. The Chief Executive Officer, Johan Ishak, commented that the contribution of creative industries in Malaysia is still low if compared to developed countries like Canada (7.4%), Great Britain (6.4%) and Singapore (5.6%). Nevertheless, he expected creative industries will contribute RM33 billion (2.2%) of the projected GDP by the year 2020 in the event that the said industry is growing at 11% annually (The Edge, 2012). In Malaysia, the National Creative Industry Policy (NCIP) was formulated in 2009 to oversee the country's creative sector which are divided into three main scopes, namely, (i) Creative Multimedia; (ii) Arts and Culture; (iii) Culture Heritage (MICC, 2010). The classifications and taxonomy of creative industries in the Malaysian context will be discussed in Chap. 4.

In terms of monetary incentives, various types of funding were provided by the Malaysian government to develop the creative industries albeit in a fragmented manner. Due to the fluidity of meaning and definitional overlap between the arts, culture, technology and the knowledge economy, thus, the availability and distribution of grants/funding tend to be sector-centric where they support either the creative arts or creative technology/entrepreneur. For example, during the 2010 Budget, then Malaysian Prime Minister announced the establishment of *Tabung Kebajikan Penggiat Seni* (Arts Welfare Fund) which is specifically for artists. Under this fund, RM3 million worth of grant was offered with initiatives to ensure the welfare of artists (Abdul Razak, 2009). Under the Tenth Malaysia Plan (2011–2015), the government established the National Creative Industry Grant with an allocation of RM120 million grant. The grant was managed by the Division of Policy and Strategic Planning (Culture), Ministry of Information, Communications and Culture (MICC) with the objective to accelerate the development of Malaysia's creative industries towards internationalisation (MICC, 2010). In 2012, the launching of MyCreative Ventures Sdn. Bhd strategised the government's investment in financing local creative companies through debt or equity investment (*What is MyCreative*, n.d.). The allocation of RM200 million from the 2012 Budget was for MyCreative to support and spur local creative entrepreneurs in the fields under NCIP's classification. According to the Malaysian Communications and Multimedia Commission (MCMC 2014) (cited in Barker & Lee, 2017: 29), another allocation of RM100 million of the Content Industry Development Fund[4] had successfully funded the completion of 44 creative content development projects and out of which 39 projects have been commercially launched by the grantees. In June 2017, the Cultural Economy Development Agency (CENDANA) was formed as an arts development organisation. Instead of the knowledge economy, CENDANA advocates the development of Malaysia's cultural economy by energising, empowering and reorganising the arts and cultural sectors through financial funding, upskilling programs and stakeholders' collaboration platforms (CENDANA, n.d.).

Besides investments in creative industries, policy-makers and urban planners are now looking into creative strategies towards urban revitalisation and regeneration (Rosly & Abdul Rashid, 2013). The public and private sector as well as non-governmental organisations have started to engage in creative initiatives with the aim to restore the vibrancy of urban streets and public spaces through arts, culture or heritage. To date, local scholars have shown interest in the area of creative cities with George Town (Penang) as case study (Khoo & Badarulzaman, 2011; Khoo et al., 2013, 2014, 2015; Rosly & Abdul Rashid, 2013; Khoo, 2016). Other urban regeneration projects that incline towards a creative city strategy are noticed in major cities like Kuala Lumpur. Recently, Johor's state government has embarked an initiative to formulate Johor's Creative Economy Blueprint (Bernama, 2019).

1.4 Why Study Creative City Development in Malaysia?

Against the above backdrop, the study of creative city development in Malaysia is timely and motivated by the following reasons. First, Malaysian cities just like their foreign counterparts are also undergoing urban restructuring following the broader shift of the nation's economic base as

[4] Content Industry Development Fund (CIDF-MCMC) programme was launched since 2011. It focuses on development of content for TV, mobile and internet. It aims to develop local creative content industry in line with the national policy objectives under the Communications and Multimedia Act 1998, which targets Malaysia as a global content development hub.

elucidated above. The macro changes are also manifested at the meso and micro levels in Malaysian cities where urban spaces and places restructure in tandem. While some Malaysian cities are losing out their original economic functions as tin ore cities (i.e. Kuala Lumpur, Ipoh), trading town (i.e. George Town) and such, there is growing interest and appeal by local urban managers to rebrand their cities as sites of cultural and creative pursuits as a way to remain relevant.

Second, the creative city terminology is cliched and too Western-centric with little mention of its suitability and applicability for cities in the East, particularly Southeast Asia. To fill this gap, this book is a novel attempt to investigate the notion of creative city development by comprehending and dissecting local nuances and idiosyncrasies of selected Malaysian cities as these cities test out this strategy in their urban development trajectory.

Third, Malaysia's ethnic plurality and cultural heterogeneity are cultural and creative endowments in their own right and they are manifested differently in each Malaysian city. Located in Southeast Asia, Malaysia has a land area of 328,550 sq. km. which comprises of two parts, namely, Peninsular Malaysia to the west and East Malaysia (Malaysia Borneo) to the east. Geographically, Peninsular Malaysia is situated south of Thailand, north of Singapore and east of Sumatera, whilst East Malaysia covers predominantly the northern part of Borneo island as seen in Fig. 1.1. East Malaysia is also bordered by Brunei to the north and Indonesian Borneo to the south. Malaysia is a unique, pluralistic and culturally diverse country in the Southeast Asian region shaped by its historical legacy. The formation of modern Malaysia can be traced back to the existence of Malay kingdoms like the Melaka Sultanate (1400s onwards), Johor kingdoms (1500s onwards) and kingdoms of other Malay states like Perak, Pahang, Negeri Sembilan, Selangor, Terengganu, Kedah and Kelantan. Such rich culture and embedded creativity in each Malaysian city warrant them excellent case studies to be examined.

Fourth, as pioneering work in the field of creative city in Malaysia, this book has great significance with scholastic and practical contributions to the agenda of creative city-making in Malaysia. Malaysia's nation-wide structural socio-economic shift following the implementation of the New Economic Policy (since 1970) has permeated into individual Malaysian states and their respective cities albeit manifested in diverse ways. The four major Malaysian cities chosen for this book, namely Kuala Lumpur, George Town, Ipoh and Johor Bahru, that have urbanised, industrialised and undergone varying degrees of urban restructuring will serve as good

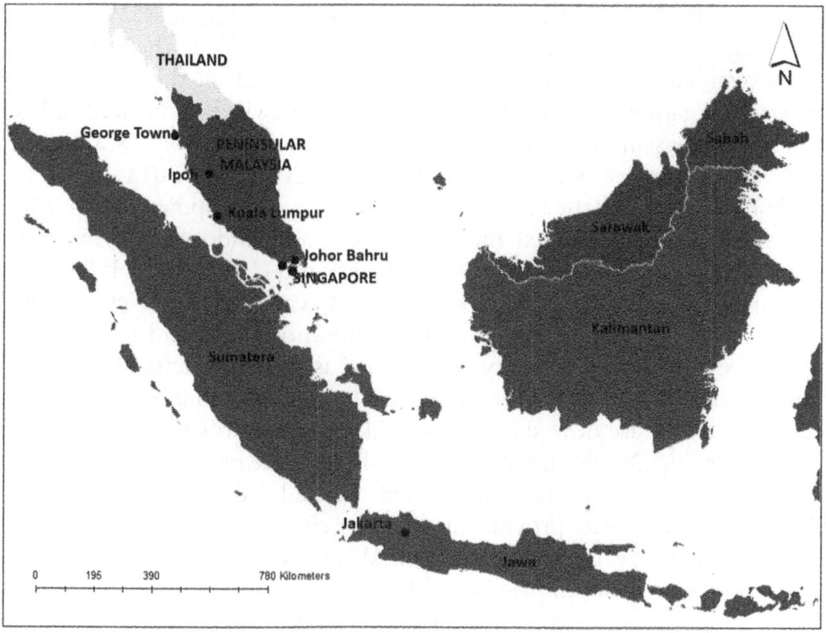

Fig. 1.1 Location map of Malaysia (showing Kuala Lumpur, George Town, Ipoh and Johor Bahru). (Source: Authors' own map, 2020)

case studies to fathom the promise and perils of creative city as a viable urban development strategy in Malaysia. In brief, the justifications for each city are as follows: (i) George Town, situated at the Northern region, is at a crossroads amidst urban conservation and gentrification; (ii) Kuala Lumpur, located right smack in the centre of Peninsular Malaysia, is wrestling to rediscover its cultural and creative prowess amidst economic globalisation and modernisation; (iii) Ipoh, a forlorn tin mining city, is rediscovering its hidden traditional as well as new cultural and creative endowments; and (iv) Johor Bahru, a border city, is gradually reaffirming itself as a creative and cultural city in Malaysia's southern region contesting alongside Singapore which is a Creative City of Design. Given that creative city formation taps into the elements of culture and creativity, the significant contribution of this book resides in the novel attempt of identifying, showcasing and utilising undiscovered local knowledge and unique cultures/creativity in Kuala Lumpur, George Town, Ipoh and Johor

Bahru to stimulate and sustain local economic development. While these are all Malaysian cities, individually, they are distinctive in their own ways. By investigating these four Malaysian cities, it will fill the research and theoretical gaps in current creative city discourse, which is still at its infancy for the Malaysian context. Despite being widely contested as a strategy that divides and exacerbates urban inequality (Pratt, 2011; McCann, 2007), nonetheless, the viability of creative city development as an urban development has garnered much attention and endorsement by urban managers in cities in the global North and South (Lee, 2011; Hidayat and Asmara, 2017). In fact, the UNESCO Creative Cities Network was founded in 2004 to foster ties and collaborations with and among cities that have earmarked creativity and culture as strategic factors for sustainable urban development. More importantly, the endeavours of the Creative Cities Network are now aligned and will move towards the 2030 Agenda for Sustainable Development where it will advocate inclusion, equality and dialogue in ensuring no one is left behind (UNESCO, 2019: 7). Additionally, creativity provides avenues for cities to develop more sustainable and inclusive economic growth. With creative sector employment now emphasising on knowledge and innovation, job creation is normally ingrained in local economies, less likely to be relocated globally and more propitious for social inclusion (UNESCO, 2019: 2).

Lastly, this book is also an attempt to decipher the types of creative city model adopted by selected Malaysian cities since there cannot be a one-size-fits-all model for all Malaysian cities. The time is opportune and apt for research in creative city development given the way globalisation has impacted and restructured each Malaysian city. In turn, these cities responded based on their respective local and contextual nuances. In the last decade especially, interest and focus towards leveraging Malaysian urban cultural assets and endowments have somewhat permeated into Malaysian urban policy arena though in a disorganised and fragmented manner. The attention given to explore the potentials of creative and cultural industries in Malaysia is disproportionate with greater emphasis in major cities like Kuala Lumpur and George Town, and emerging interest in secondary cities like Johor Bahru and Ipoh. Although the label 'creative city' is largely absent and not formalised in most of Malaysia's city development plans except for Kuala Lumpur with the recent '2019 Kuala Lumpur Creative and Cultural District Strategic Master Plan', local initiatives in George Town, Johor Bahru and Ipoh have started to acknowledge and harness creative and cultural industries to revitalise their cities. The subsequent concern would be the viability and sustainability of the creative city model adopted by each city; whether it is organically developed, supply-driven or a novel reconceptualisation of a Malaysian creative city model to its own right.

References

Abdul Razak, N. (2009, October 23). The 2010 Budget Speech by Prime Minister & Minister of Finance. *Introducing the Supply Bill (2010)*. The Dewan Rakyat.

Aghion, P., & Bircan, C. (2017). *The Middle-Income Trap from a Schumpeterian Perspective* (ADB Economics Working Paper Series 521).

Andersson, A. E. (2011). Creative People Need Creative Cities. In D. E. Andersson, A. E. Andersson, & C. Mellander (Eds.), *Handbook of Creative Cities* (pp. 14–55). Edward Elgar Publishing Limited.

Barker, T., & Lee, Y. B. (2017). Making Creative Industries Policy: The Malaysian Case. *Kajian Malaysia, 35*(2), 21–37.

Bernama. (2019, February 26). Johor Identifies JB, Ayer Hitam and Parit Sulong for Possible UNESCO Listing. *New Straits Times*, Malaysia.

Boccella, N., & Salerno, I. (2016, May 18–20). *Creative Economy, Cultural Industries and Local Development*. The 2nd International Symposium 'New Metropolitan Perspectives' – Strategic Planning, Spatial Planning, Economic Programs and Decision Support Tools, Through the Implementation of Horizon/Europe2020. ISTH2020, Reggio Calabria, Italy.

CENDANA. (n.d.) *Year One in Review*. Retrieved from: https://www.cendana.com.my/about-us/year-one-in-review/year-one-in-review. Accessed 20 Sep 2019.

Chaudhury, S., & Lundbery, A. (2018). Singapore as a Creative City: Vignettes from the Perspective of la flaneuse tropique in eTropic 17.2. *Tropical Imaginaries in Living Cities*, Special Issue, 110–131.

Cornell University, INSEAD, & WIPO. (2016). *The Global Innovation Index 2016: Winning with Global Innovation*. Ithaca, Fontainebleau.

Digital News Asia. (2014, March 19). *Malaysia's Creative Content Industry Worth Nearly US$5 Bil, Says MDeC*. Retrieved from: https://www.digitalnewsasia.com/node/6661. Accessed 20 Sep 2019.

DoSM (Department of Statistics, Malaysia). (2020). *Household Income and Basic Amenities Survey Report 2019*. Department of Statistics, Malaysia.

Florida, R. (2002). *The Rise of the Creative Class: And How It's Transforming Work, Leisure, Community and Everyday Life*. Basic Books.

Florida, R. (2004). *Cities and the Creative Class*. Routledge.

Florida, R. (2007). *The Flight of the Creative Class: The New Global Competition for Talent*. HarperCollins Publishers.

Florida, R. (2008). *Who's Your City? How the Creative Economy Is Making Where to Live the Most Important Decision of Your Life*. Basic Books.

Galloway, S., & Dunlop, S. (2007). A Critique of Definitions of the Cultural and Creative Industries in Public Policy. *International Journal of Cultural Policy, 13*(1), 17–31.

Hasnul, N. S. (2011, October). *A Snapshot of Malaysia's Creative Multimedia Industries: Trends and Opportunities*. Seminar on 'Enhancing the Competitiveness of the Creative Industry', MATRADE, Malaysia.

Hidayat, AR. R. T., & Asmara, A.Y. (2017). *Creative Industry in Supporting Economy Growth in Indonesia: Perspective of Regional Innovation System*. The 3rd International Conference of Planning in the Era of Uncertainty, IOP Conference Series: Earth and Environmental Science 70.

Hisham, H. T. (2016, May 25). *Sumbangan industri kreatif kepada negara semaking meningkat*. Bacalah Malaysia. Retrieved from: http://www.bacalahmalaysia.com/2016/05/25/sumbangan-industri-kreatif-kepada-negara-semakin-meningkat/. Accessed 13 June 2017.

Jacobs, J. (1972). *Economy of Cities*. Penguin Books.

Karvelyte, K. (2018). Shifting Meanings in Changing Contexts: The Role of the Creative City in Shanghai, Hong Kong and Taipei. *International Journal of Cultural Policy*. https://doi.org/10.1080/10286632.2018.1479748

Khoo, S. L. (2016). Understanding Culture and Creativity in Urban Development: What It Means for George Town, Penang? *Geografia Malaysia Journal of Society and Space, 12*(13), 1–11.

Khoo, S. L., & Badarulzaman, N. (2011, December 6–7). *Planning George Town as a Creative City: A Conceptual Framework*. The 5th Conference on 'Built Environment in Developing Countries – Rehumanizing the Built Environment', Vistana Hotel, Penang, Malaysia.

Khoo, S. L., Badarulzaman, N., Samat, N., & Dawood, S. R. S. (2014). Capitalising on Urban Cultural Resources for Creative City Development: A Conceptual Review and the Way Forward for Malaysia's George Town. *Geografia Online Malaysian Journal of Society and Space, 10*(5), 20–29.

Khoo, S. L., Samat, N., Badarulzaman, N., & Dawood, S. R. S. (2015). The Promise and Perils of the Island City of George Town (Penang) as a Creative City. *Urban Island Studies, 1*(1), 20–34.

Khoo, S. L., Samat, N., Badarulzaman, N., Sirat, M., & Dawood, S. R. S. (2013). Spatial Mapping of Creative Industries in Creative Cities: The Case of Penang, Malaysia. *Humanities and Social Sciences Review, 2*(2), 151–169.

KRI (Khazanah Research Institute). (2017, October 10). *An Uneven Future? An Exploration of the Future of Work in Malaysia* (Discussion Paper 2/17). Khazanah Research Institute.

KRI (Khazanah Research Institute). (2018). *The State of Households 2018: Different Realities* (3rd ed.). Khazanah Research Institute.

Lafaye de Micheaux, E. (2017). *The Development of Malaysian Capitalism: From British Rule to the Present Day*. Strategic Information and Research Development Centre.

Landry, C. (2005). Lineages of the Creative City. In S. Franke & E. Verhhage (Eds.), *Creativity and the City: How the Creative Economy Is Changing the City*. NAI Publishers.

Landry, C. (2008). *The Creative City: A Toolkit for Urban Innovators* (2nd ed.). Comedia Earthscan.

Landry, C., & Bianchini, F. (1998). *The Creative City*. Demos.

Lee, C. B. (2011). Creative Economies, Creative Cities. *International Journal of Cultural Policy, 17*(3), 356–360.

Lim, G. E. (2014, May 6). *Penang Chief Minister's Speech*. The Opening of Penang Creative Industry Conference & Showcase PCICS, Equatorial Hotel, Penang, Malaysia.

MATRADE (Malaysia External Trade Development Corporation). (2017). *Boosting the Exports of Malaysia's Creative Content Through Partnership*. Retrieved from: http://www.matrade.gov.my/en/about-matrade/media/press-releases/3735-boosting-the-exports-of-malaysia-s-creative-content-through-partnership. Accessed 20 Sep 2019.

McCann, E. J. (2007). Inequality and Politics in the Creative City-Region: Questions of Livability and State Strategy. *International Journal of Urban and Regional Research, 31*(1), 188–196.

Mercer, C. (2006). *Cultural Planning for Urban Development and Creative Cities*. Publisher Unknown.

MICC (Ministry of Information, Communications and Culture). (2010). *National Creative Industry Policy (NCIP)*. Ministry of Information, Communication and Culture.

Pratt, A. (2011). The Cultural Contradictions of the Creative City. *City, Culture and Society, 2*, 123–130.

Rosly, D., & Abdul Rashid, A. (2013, April 3–6). *Revitalising Urban Development in Malaysia Through the Implementation of Urban Regeneration Programme*. The 43rd Annual Conference of the Urban Affairs Association, San Francisco California.

Smidt-Jensen, S. (2007, January). *The Roles of Culture and Creativity Within Urban Development Strategies: Outlining a Theoretical Framework for Analysing Scandinavian Cities* (CSB Working Paper No. 8). Centre for Strategic Urban Research.

TCPD (Town and Country Planning Department). (2016). *Second National Urbanisation Policy (NUP2)*. Town and Country Planning Department.

The Edge. (2012, September 13). *GDP Contribution from Creative Industry to Reach RM33b by 2020*. The Edge Markets. Retrieved from: https://www.theedgemarkets.com/article/gdp-contribution-creative-industry-reach-rm33b-2020-0. Accessed 16 Sep 2019.

UNCTAD (United Nations Conference on Trade and Development). (2017). *Strengthening the Creative Industries for Development in the Republic of Korea*. United Nations.

UNESCO. (2019). *Voices of the City: UNESCO Creative Cities Moving Towards the 2030 Agenda for Sustainable Development*. UNESCO.

What Is MyCreative. (n.d.). Retrieved from MyCreative Ventures Sdn. Bhd. website: http://www.mycreative.com.my/about. Accessed 20 Sep 2019.

World Bank. (2017). Turmoil to Transformation: 20 Years After the Asian Financial Crisis. In *Malaysia Economic Monitor*. World Bank.

CHAPTER 2

Creative City Discourse

Abstract This chapter will broadly illustrate the origins of the creative city concept alongside its highly debatable conceptualisation in the discourse. As the dialectics are compounded by spatial, temporal and contextual variations where the hype of creative city-making has transcended borders from Western to Eastern shores, this chapter will chronicle the origins and unravel the underlying agendas behind creative city development in the Western world. This chapter will also discuss the converging or diverging pathways by Eastern creative cities.

Keywords Creative city discourse • Western lens • Eastern lens • Convergence • Divergence

2.1 Early Definitions from the West

The concept of creative city is not novel. Despite garnering much attention among scholars (Pratt, 2008; Oliveira & Paulino, 2017; Grodach, 2017; Landry, 2005, 2008) and policy-makers in the last few decades (European Commission, 2017; AuthentiCity, 2008), the conceptualisation of creative city dates way back in time with diverse connotations dictated by semantics, functionality as well as practicality. Underpinned by elements of culture, creativity and the dynamism of urbanisation, the creative city concept developed predominantly from Western lenses that identify the functionality of creative cities as informal and unconstrained

S. L. Khoo, N. S. F. Chang, *Creative City as an Urban Development Strategy*, https://doi.org/10.1007/978-981-16-1291-6_2

changing spatial organisations for huge creative revolutions. For the past 2500 years, scholars have identified relatively big Western (creative) cities, albeit from different eras. Examples are Athens at 400 BC, Renaissance Florence, Enlightenment London to name a few, and they functioned as cradles of revolutionary creativity. As surmised by Andersson (2011: 14), creative cities not only provide a conducive environ to cultivate creativity among existing urban dwellers but the city also functions like a 'magnet' to attract migrants with creative dispositions. With the clustering of creative people in the same location, creative cities are indeed fecund localities to grow, create and innovate new ideas in alliance with other creative talents (Andersson et al. 1984) (cited in Andersson, 2011: 14). Although the coming of the post-industrial society (Bell, Toffler, etc.) emphasised the element of creativity, there is indeed flow, linkage and intertextuality between past creative cities and contemporary creative cities in today's globalised era. Undoubtedly, creative cities are and have unfailingly been linked with diverse and massive influx of migrants as well as flow and exchange of goods and services coupled with new artistic and scientific ideas (Andersson, 2011).

Broadly, the discourse on creative cities development can be understood from three diverse yet interlinked disciplines, namely, (i) regional economics; (ii) urban planning; and (iii) psychology (Andersson & Mellander, 2011: 4). However, for the purpose of this book, the focus will be on Malaysian creative city-making from an urban planning, regional development and development studies perspectives. Contemporary popularity and debates on creative cities development have departed from Florida's famed publication 'The Rise of the Creative Class', which was in fact an offshoot from seminal works by Joseph Schumpeter, Peter Drucker, and Daniel Bell who prophesised the rise of a knowledge-based economy that leverages on knowledge, creativity and innovation. According to Florida, we now enter the creative ethos amidst ongoing changes around us. Echoing and endorsing Schumpeter, Drucker and Bell's convictions of the coming of a post-industrial society, Florida further articulated that we now live in an era where the key determinant of economic life resides in 'human creativity' – a valuable element where systems have transformed to advocate and leverage it. Creativity, arguably, will give rise to new technologies, new industries, new wealth and other favourable economic aspects (Florida, 2004: 21). Florida also extended from Paul Romer's New Growth Theory economic school of thought by arguing that the uniqueness of humans is not merely creativity present in each human being; but it is such creative prowess that economically differentiates humans from other species. Romer perceived creativity as economic goods that do not deplete but can be used repeatedly and increase in value as it is used more frequently, thus, providing increasing returns instead of diminishing returns. In sum, creativity as a generator of economic value (Florida, 2004: 37).

Broadly, Florida's depiction of a creative city culminates in a term he coined – creative class – which is actually an extension of the type of knowledge workers required in Peter Drucker's 'knowledge economy' and creative employees sought after in John Howkins' 'Creative Economy'. Nonetheless, the creative class has a differentiating attribute where its fraternity involves in work with the function to 'create meaningful new forms' (Florida, 2004: 68). Florida's creative class comprises of two categories, namely, (i) 'Super Creative Core' who are involved in creative processes like scientists, designers, architects, engineers, academics, poets, novelists, artists, actors, writers, think-tank researchers, editors, analysts; (ii) 'Creative professionals' such as those hired in knowledge-intensive industries like financial services, healthcare, legal, business management and high-tech sectors. Creative professionals are expected to think on their own, tweak standard procedures to address new circumstances, make informed judgements and be capable of solving problem by tapping on complicated corpus of knowledge (Florida, 2004: 69).

Arguably, many of the creative city notions especially Florida's viewpoints, depart from economic geography and regional science by advocating agglomeration economy which is geographical clustering of firms that collaborate (or even compete) and eventually contribute to regional growth. Again, these are not novel discoveries but an extension from previous works by seminal theorist Alfred Marshall (1966) who popularised the importance of 'co-locating'. He perceived cities as 'clusters' and recognised the benefits of agglomeration of similar firms/factories/ industries and localised industries as the avenue to conceive new ideas, creations and innovations (Marshall, 1966: 225). As a result, this brought about positive externalities linked to 'specialisation of the labour market' where close proximity of firms, industries and human capital within a locality is advantageous and yield benefits in spreading expertise, knowledge and spillover effects (Lorenzen & Frederiksen, 2008: 158).

Similarly, when viewed from a broader regional perspective, early works by Western scholars such as Walter Isard (1956) have popularised the agglomeration economies concept by canvassing the link between economic growth, industrialisation, modernisation and urban development. Isard argued that cities emerged due to agglomeration economies where city formation purportedly generates costs and benefits to producers and consumers. He categorised agglomeration economies into two, namely, (i) localisation economies, and (ii) urbanisation economies. Localisation economies is somewhat similar to the notions of clustering, agglomeration and co-locating where 'agglomeration effect' is generated when certain economic sectors grow within a location (Todaro & Smith, 2011: 318). As for urbanisation economies, the concept harnesses the importance of 'place'. The promise of urban localities (i.e. modern amenities, greater employment opportunities, etc.) has influenced agglomeration effects with the overall growth of a concentrated economic region. Purportedly, urbanisation economies leverage on a spectrum of ideographic and place-specific factors that mainly portray urban attributes (Lorenzen & Frederiksen, 2008: 159; Florida, 2002, 2008). Generally, urbanisation economies are divided into three (3) categories. First, the co-location and agglomeration of different firms thus resulting in a city's diverse industry (Ohlin 1993) (cited in Lorenzen & Frederiksen, 2008: 159). Second, a city's diverse labour force and its heterogeneity. This notion was popularised by Jane Jacobs in the 1970s where cities were viewed as a symbiotic co-existence of multiple layers of clusters, constant immigration of labour, intersected labour force and a rich surrounding. Subsequently, Jacobs argues that this rich urban setting will then condition the birth of novel ideas, new skills, know-how and expertise that instil creativity and innovation in individuals, firms and industries (Jacobs, 1972). The third urbanisation economies category acknowledges the infrastructural and institutional variety of a place. Undeniably, easy accessibility of a spectrum of infrastructures are fundamental to develop an appealing urban environ that attracts diverse labour to migrate to the city.

Arguably, scholars have identified one common goal that creative cities share which is their attractiveness to enterprises, investors, knowledge workers and visitors and this is a situation that echoes Harvey's 'entrepreneurial cities' (Harvey 1989) (cited in Karvelyte, 2018). Due to increasing competition among cities, Harvey argued that entrepreneurial cities expanded urban entrepreneurialism to lure investors by short-term 'civic boosterism' practices. Apart from viewing the creative city concept from an economic and entrepreneurial perspective, scholars have also identified another orientation that fleshes out the cultural aspect of cities (Landry, 2005, 2008; Landry & Bianchini, 1998). However, fuzzy conceptualisation of what constitutes culture, creativity and cultural/ creative industries will make it challenging to ascertain the fundamentals that underscore a culturally based creative city. In the Western world, the contentions are such that cities were hollowed out. In response to globalisation, urban restructuring occurred with an option to shift to the services and creative economy. To this end, Western proponents have identified a city's 'culture' as the new factor of production to drive a city's economic development (Landry, 2005, 2008; Landry & Bianchini, 1998). Correspondingly, the place-specific and location-uniqueness of certain cultural assets (i.e. historic structures, local knowledge, cultural heritage) embedded within a place have spurred most Western urban managers to thread along the creative city pathways (i.e. London, Liverpool, Manchester, Toronto), though without clear conviction whether they understand the benefits and disbenefits of this strategy.

The strategy of capitalising culture was a big bang concept in Europe beginning from 1980s when the European Capitals of Culture initiative (1985–2010) kicked off. European cities like Athens (1985), Florence (1986), Amsterdam (1987), Berlin (1988), Antwerp (1993), Copenhagen (1996), Stockholm (1998), Bologna (2000), Brussels (2000), Rotterdam (2001), Salamanca (2002), Liverpool (2008) Essen (2010), Istanbul (2010), Talinn (2011) and such have started to identify themselves as cultural capitals in Europe. By then, the UNESCO Creative Cities Network was established in 2004 to advocate collaboration with and among cities that leverage creativity as a strategic factor for sustainable urban development. With 246 member cities to date, the network strives towards a common goal of positioning creativity and cultural industries at the crux of their development plans locally and at the international platform (UNESCO Creative Cities Network, 2019).

In sum, scholars have delineated creative city orientations into two, namely, an econo-centric or culture-centric orientation. While the econo-centric viewpoint tends to capitalise on economic sustainability, creativity and innovation akin to the knowledge-based mantra, the culture-centric lenses lean towards cultural and artistic endeavours of a creative city as a social inclusion strategy (Smith & Warfield, 2008: 289).

In retrospect, the terminology 'creative industries' was intrinsically linked to 'cultural industries' with origins from scholars like Horkheimer and Adorno (2002) (cited in Galloway & Dunlop, 2007: 18) who defined cultural industries as those industrially produced commercial entertainments (i.e. broadcasting, film, publishing, recorded music) as being different from subsidised 'arts' (i.e. visual and performing arts, museums and galleries). This depiction underscored UNESCO and European cultural industries policy in the 1980s. Subsequently, societal and developmental shift saw the rise of the knowledge economy which introduced information and communication technology to all realms of the economy where the cultural industries are no exception. Following suit, political leaderships in Australia and the UK made a 'decisive shift in terminology' where the term cultural industries slowly and eventually dissipated and was substituted by 'creative industries' in policy realms (Galloway & Dunlop, 2007: 18). In Australia, the term was first coined in a report produced for the Australian government in the early 1990s to tabulate economic activities. In 1994, the Creative Nation report was released by then Prime Minister Paul Keating where a cultural policy was inaugurally formulated by the Australian federal government. An additional funding of AUD250 million was allocated to cultural institutions across Australia.[1]

Later in the United Kingdom, the creative industries concept emerged with the election of the New Labour government in 1997 and the creation of the Department of Culture, Media and Sports (DCMS) (O'Connor, 2010). However, the origin of the terminology 'creative industries' itself is slippery. Scholars argued that the term derived from semantic adjustment when culture and cultural industries have been carelessly subsumed within the broader creative industries agenda under the overarching knowledge economy (Galloway & Dunlop, 2007: 25). Such manner of re-classifying creative industries has been heavily critiqued and will be further espoused in Chap. 3. In the UK, by setting up their Creative Industries Task Force, the DCMS produced the landmark Creative Industries Mapping Document in 1998 to systematically define and measure creative industries. The DCMS defined creative industries as: 'those activities which have their origin in individual creativity, skill and talent and which have a potential for wealth and job creation through the generation and exploitation of intellectual property'. Initially, DCMS started with a 13-sector classification comprising of (1) advertising, (2) architecture, (3) arts and antique markets, (4) crafts, (5) design, (6) designer fashion, (7) film and video, (8) interactive leisure software, (9) music, (10) performing arts, (11) publishing, (12) software and computer services, and (13) television and radio (DCMS, 2010). Subsequently, the latest statistical release by the DCMS (2015) has reorganised these sectors into nine (9)

[1] Creative nation: Commonwealth cultural policy, October 1994, Department of Communications and the Arts (Australia), 30 October 1994, Source: http://apo.org.au/node/29704 (accessed on 22 May 2018).

categories as follows: (1) advertising and marketing, (2) architecture, (3) crafts, (4) design (product, graphic and fashion design), (5) film, TV, video, radio and photography, (6) IT, software and computer services, (7) publishing, (8) museums, galleries and libraries, and (9) music, and performing and visual arts. At this point, there is clear difference between the definitions by DCMS and UNESCO. The DCMS definition emphasises innovation, intellectual property and wealth creation (akin to the mantras in a knowledge-based creative city model) whereas UNESCO's depiction of creative industries leans towards culture with emphasis on cultural and artistic content whilst incorporating new information and communication technologies to advance innovation as quoted below.

> Cultural and creative industries are those sectors that have as their main objective the creation, production, distribution and consumption of goods, services and activities that have cultural and artistic content. They are characterised by being at the intersection of economy and culture, having creativity at the core of their activities, artistic and/or cultural content, and links to innovation. (Habitat III, 2015: 1)

Subsequently, the definition of creative cities by Habitat III is as follows:

> Creative cities are defined as urban complexes where cultural activities are an integral component of the city's economic and social functioning, for example through support to cultural and creative professionals, enhanced investments in cultural infrastructure, creative industries and new ICTs, or the adoption of bottom-up approaches to urban development. (Habitat III, 2015: 1)

Efforts to map and measure the cultural industries were also undertaken by cultural economists like Throsby (2008) where he managed to position culture as well as depict the structural characteristics of the cultural industries and arts within the broader economy. Acknowledging the fact that creative ideas generate cultural content in cultural industries, he argues that cultural goods and services do produce two different types of values, namely, economic and cultural (Throsby, 2008: 148, 150). However, there are also scholars who cautioned against the careless way of defining and delineating cultural and creative industries by conceptualising and subsuming cultural industries under the knowledge-based and information society discourse (Kong, Halloway, etc.). These critiques will be discussed further in Chap. 3.

2.2 Fathoming Creative City Discourse from Eastern Lenses

In the twenty-first century, the hype of creative city-making has gradually transcended Western borders to Eastern shores. Eastern cities adopted this Western-origin concept rather differently. Broadly, Eastern cities lean towards the culture-centric approach with a focus on local arts and cultural identities, and with stronger state presence and intervention (Karvelyte, 2018: 2). The adoption of the creative city concept within Eastern shores are actively being broadcasted through the UNESCO Creative Cities Network (UCCN). Asian cities like Singapore and cities in Japan, China, Thailand, Indonesia, Korea, India and Vietnam are members of the network. The following sections depict some case studies of creative city-making in Asian cities.

In 2005, UNESCO recognised China as the leading Asian nation in development of creative and cultural economy in the East. In 2008, UNCTAD's report reveals China as the world's biggest exporter of creative goods. The creative industries contributed almost 2.5% of China's GDP and generated 11.32 million employments in 2006 (Kern et al., 2011). The creative sectors that grew significantly were online games, new media and advertising; TV, radio and publishing; film and video; design and architecture; and visual arts, crafts and cultural heritage. Shanghai and Beijing are among the earliest cities that attracted creative clustering. Since 2004, creative industry parks/clusters have mushroomed at different provinces/cities and gradually transformed China's cultural landscape. In 2008, Shenzhen was bestowed UNESCO's City of Design. The Chinese state has intervened to drive incubation and development of creative and cultural industries in China. Concurrently, regional and municipal governments played active roles in building culture-oriented development strategies (Kern et al., 2011; Fung, 2016). While driving the growth of creative industries, the Chinese government must deal with contradictions between creative content and existing censorship, control mechanism and bureaucracy of the authoritarian regime within the country. The controversy revolved around the relationship between ideological control, state competence and globalisation of local creativity and culture (Fung, 2016).

In Japanese cities, the creative city-making endeavour is actively driven by the city's government by adopting a more people-oriented strategy that is underscored by community empowerment, social inclusion and cultural protection besides economic dividends. In Japan, scholars like Sasaki led the promotion of creative city concept both in theory and practice. He proposed a 'cultural mode of production' model that serves as a well-balanced system of cultural production and consumption, leveraging cultural capital, towards producing products and services with high economic and cultural value (Sasaki, 2007, 2010). Sasaki (2010) believes that creative arts and cultural initiatives by artists, creators or ordinary citizens and socially disadvantaged groups could consequently foster social inclusiveness and empowerment. Subsequently, it improves inequality in society, and ultimately regenerates the city's development. In 2008, the local emphasis on design for product manufacturing, community's lifestyle enhancement, human capital development and cityscape development in Nagoya and Kobe have accoladed both cities to be the first two UNESCO Creative Cities in Japan and Asia. Later, under the mayor-led administrative and private efforts, Kanazawa was listed as UNESCO Creative City of Crafts and Folk Art in 2009 (UNESCO Creative Cities Network website). The implementation of the '*Monozukuri* (craftsmanship or art of manufacturing)' Ordinance to protect and promote traditional arts and crafts, and other new industries have preserved the solid foundation of Kanazawa's societies towards maintaining a high quality of life and balance between culture and economy (Sasaki, 2010). The art and cultural-led policies and initiatives in Japan have consequently caused the rise of socially inclusive politics throughout Japanese cities (Sasaki, 2010). Subsequently, Japan's government started to invest in the creative sector under the motto of 'Cool Japan' (ASEF, 2014).

Although Singapore received the UNESCO Creative City of Design accolade in 2015, Singapore's government had actively invested in creative industries much earlier, focusing on art, media and design. Statistics showed a 3.2% contribution from creative industries to Singapore's GNP in 2002 (MITA, 2002). Originality and entrepreneurship are increasingly being valued in Singapore. Since 2001, Singapore's government established the Economic Review Sub-Committee Workgroup on Creative Industries (ERC-CI) to strategically drive arts, innovation-fuelled businesses and technology as well as harness the multi-dimensional creativity of people for new competitive advantages (ERC-CI, 2002). Creative city-making in Singapore are guided by three master plans, namely, (i) Renaissance City Plan (1999), (ii) Media21 Plan (2003) and (iii) Design Singapore (2003) (Kong, 2012). Similar to China, the constrained freedom of creative expression and production due to local censorship always challenge Singapore's creative sectors (Ooi, 2008; Kong, 2012).

Indonesia is another sterling example where the government officially initiated policies to develop creative industries since 2009 under an enactment. It was then supported and driven by the Ministry of Tourism and Creative Economy in 2011 (Hidayat and Asmara, 2017). Statistics showed that Indonesia's creative sector has contributed 7% or $US46,930 billion to Indonesia's GDP and created about 11.8 million jobs in 2013, mostly in craft and fashion industries (Statistics Indonesia (BPS), 2013). In 2015, the government legitimated Creative Economy Agency *(Bekraf)* to promote and boost creative entrepreneurs in Indonesia's many regions. It focuses on film/movie; animation and design; music; craft; fashion; and culinary. At the local level, local governments across Indonesian cities are actively developing local art and cultural industries to focus on cultural consumption and production (Sawangchot, 2016).

Instead of focusing on cultural-led creative economy, Korea's government strives to integrate people's imagination and creativity to science, technology and ICT. This started since 2013 towards building new Korean industries and markets. Korea's Ministry of Science, ICT and Future Planning plays a major role in coordinating creative economy policies under the Creative Economy Policy Bureau. The government has two creative economy blueprints (i.e. Creative Economy Action Plan, Measures to Establish a Creative Economic Ecosystem) to generate new employment and markets, strengthen global leadership and create a society that respects and manifests creativity. Under the more liberal top-down approach, with support from public-private partnerships, more open creative ecosystems have developed across Korean cities. UNCTAD (2017) reported that Korea has achieved the top rank on the EU Innovation Union Scoreboard Index for three consecutive years (2014–2016). According to INSEAD and WIPO (2017), Korea achieved second among the most innovative Asian countries and the eleventh among global nations. It maintained its top position in patenting and other intellectual property-related markets. The designs, games and contents industries, especially the Korean entertainment and popular culture, contribute significantly to the national economy through exports of goods/products, contents as well as services in design and culture (UNCTAD, 2017).

In Taiwan, Taipei's creative city policy evolved since 1991 in response to different urban developmental stages. Initially, without mention of creative city concept, Taipei's city government transplanted the concept of '*Machizukuri*' (a people's community movement for community improvement started in Japan) (Watanabe, 2012) to launch local cultural industry strategy towards enhancing community participation in promoting place-specific cultural heritage sites (Lin & Hsing, 2009). The 'creative city' and 'creative class' concepts were then introduced and developed in Taipei through the city's policy-learning and knowledge transfer initiatives from global experts between 2000 and 2008. Later, the creative city concept was adopted as Taipei's urban regeneration strategy during this timeframe. Instead of aspiring to join the UNESCO Creative Cities Network, Taipei's government was proactive to develop a localised creative city model by integrating trans-local experience with local planning practices (Lin, 2016). A strong neoliberalism influence that emphasises on creative-led, consumption-oriented economic development had consequently fostered competitive urban entrepreneurialism in Taipei.

Overall, Asian creative/cultural city-making models reflect the 'global city' ambitions of Asian cities which is driven under state authoritarianism (Oakes & Wang, 2016). Due to avidity for 'worlding' as globally significant creative or cultural hubs, the cultural policies in many Asian countries have shifted far beyond the objective of promoting local culture. The branding-oriented appropriations of culture in the aspirational urbanism of Asian cities tend to neglect the local cultural content or the complex social formations. Local governments in many Asian cities (except Japanese cities) tend to overemphasise the economic ordering of cognitive-cultural capitalism (Peck, 2002), especially cultural consumption (Pratt, 2009). Fast policy transfer and symbolic place-making initiatives that fail to consider place-specific contexts and local rationales in Asian creative cities inevitably generate social-economic frictions (Peck, 2012), cultural production-consumption tensions (Sasaki, 2010), social exclusion (Catungal & Leslie, 2009), political tensions and local controversies among stakeholders (Oakes & Wang, 2016). Therefore, Asian cities should look into the strength of the city as a cultural policy generator instead of just a cultural policy recipient (Oakes & Wang, 2016). Mindful cultural economic planning practices and institutional innovations are crucial in responding to local tensions/needs (Scott, 2014; Lin, 2016).

2.3 Western Versus Eastern Lenses: Convergence or Divergence?

Fathoming concepts and reconceptualising new ones will rely on socio-temporal contexts and local nuances. While the creative city concept might be generically applied to an extent of being a cliché in urban development discourse, there are clear differentiations and some mild overlaps of definitions, classifications and taxonomies between what is fathomed as Western and East lenses. Arguably, as a concept that is underscored by 'culture' and 'creativity', it is pointless to state a claim that the concept originated from the East or West given that urban development and the extent upon which culture and creativity are cultivated have existed since time immemorial in both civilisations.

However, a glance into the chronological discourse of creative city development does point to the fact that leveraging, producing and reproducing culture and creativity to spur local urban development has Western origins. From the times of ancient Greece right till the birth of modern urban planning discourse by Peter Hall, Jane Jacobs and Charles Landry, elements of creativity, innovation and culture have gradually permeated and assumed centre stage in Western academic parlance and policy rhetoric despite varying degrees of comprehension and adoption. The presumption that a city has cultural assets that can be harnessed and the fact that the creative city mantra leans more towards a services-based, knowledge economy following the worldviews of Drucker, Toffler, Bell and later Florida, have now caught up in Eastern economies and cities. This is observable through the voluminous literature on creative cities written about Western cities since the 1980s and, undeniably, the interest has gradually traversed boundaries and permeated to Eastern shores. Hence, this is unquestionably a period where many Eastern cities are catching up and experimenting the creative city concept for their respective cities though many are not fully familiar with the concept, latent challenges and issues embedded within it. Thus, this caused some cities to embark on this strategy with the role of 'follower' rather than questioning its viability and applicability to the local context.

Arguably, the point of divergence with regard to creative city development is the way culture, creativity and commodity are understood, utilised, commodified and reproduced in both Western and Eastern urban milieus. While Western creative city development predominantly showcases culture and creativity embedded in products, services and experiences that are Western-centric with little or no infiltration from outside cultures or creativity models, the same cannot be said about creative city development in Eastern cities. Though Eastern cities are rich in their respective cultural endowments and creativity, some cities in the East tend to overlook and disregard their own native and original cultural assets and instead glorify foreign and Western cultures that are alienated to the local context.

So far, based on existing discourse, the Eastern interpretation and adoption of creative city concept tend to be more culture-centric instead of econo-centric. The Eastern focus leans towards the artistic and cultural economy, leveraging the place-specific or location-uniqueness of the art, cultural and heritage assets found and inherited among urban communities. It is very much influenced or motivated by the UNESCO Creative Cities Network (UCCN) where cities are getting connected and sharing experiences globally through creative city-making endeavours in promoting creativity as a driver for sustainable urban development through economic regeneration, social inclusion and cultural vibrancy. Under many circumstances, top-down instead of bottom-up approaches are applied in driving creative city strategies in Eastern nations while the readiness and tolerance towards the openness and acceptance of creativity among Eastern communities are still lacking if compared to Western communities. This is among the key challenges for the culture-centric approach in creative city-making in most Eastern countries, mainly due to their inherited cultural backgrounds, low levels of awareness and appreciation towards creativity under different educational systems and emphasis. Consequently, that generates conflicts and controversies among the creative producers, policy-makers and different groups of stakeholders.

References

Andersson, A. E. (2011). Creative People Need Creative Cities. In D. E. Andersson, A. E. Andersson, & C. Mellander (Eds.), *Handbook of Creative Cities* (pp. 14–55). Edward Elgar Publishing Limited.

Andersson, A. E., & Mellander, C. (2011). Analysing Creative Cities. In D. E. Andersson, A. E. Andersson, & C. Mellander (Eds.), *Handbook of Creative Cities* (pp. 3–13). Edward Elgar Publishing Limited.

ASEF. (2014). *Enabling Crossovers: Good Practices in the Creative Industries.* Asia-Europe Foundation (ASEF), June 2014. Retrieved from: https://www.asef.org/images/ASEF_Publication_EnablingCrossovers.pdf. Accessed 20 Sep 2019.

AuthentiCity. (2008). Creative City Planning Framework, A Supporting Document to the *Agenda for Prosperity: Prospectus for a Great City* (Prepared for the City of Toronto). City of Toronto.
Catungal, J. P., & Leslie, D. (2009). Placing Power in the Creative City: Governmentalities and Subjectivities in Liberty Village, Toronto. *Environment and Planning A, 41*(11), 2576–2594.
DCMS (Department for Culture, Media and Sport). (2010). *Creative Industries Economic Estimates (Experimental Statistics)*, Full statistical release, December 9, 2010. Retrieved from: http://www.culture.gov.uk/images/research/CIEE_Full_Release_Dec2010.pdf. Accessed 10 Sep 2019.
DCMS (Department for Culture, Media and Sport). (2015). *Creative Industries Economic Estimates – January 2015*, Statistical release. Retrieved from: https://www.gov.uk/government/uploads/system/uploads/attachment_data/file/394668/Creative_Industries_Economic_Estimates_-_January_2015.pdf. Accessed 10 Sep 2019.
ERC-CI. (2002). *Creative Industries Development Strategy: Propelling Singapore's Creative Economy.* Economic Review Committee – Services Subcommittee Workgroup on Creative Industries (ERC-CI).
European Commission. (2017). *The Cultural and Creative Cities Monitor.* European Commission Joint Research Centre.
Florida, R. (2002). *The Rise of the Creative Class: And How It's Transforming Work, Leisure, Community and Everyday Life.* Basic Books.
Florida, R. (2004). *The Rise of the Creative Class.* Basic Books.
Florida, R. (2008). *Who's Your City? How the Creative Economy Is Making Where to Live the Most Important Decision of Your Life.* Basic Books.
Fung, A. (2016). Strategizing for Creative Industries in China: Contradictions and Tension in Nation Branding. *International Journal of Communication, 10*, 3004–3021.
Galloway, S., & Dunlop, S. (2007). A Critique of Definitions of the Cultural and Creative Industries in Public Policy. *International Journal of Cultural Policy, 13*(1), 17–31.
Grodach, C. (2017). Urban Cultural Policy and Creative City Making. *Cities, 68*, 82–91.
Habitat, I. I. I. (2015). *Habitat III Issue Papers, 4 – Urban Culture and Heritage.* United Nations Conference on Housing and Sustainable Urban Development.
Hidayat, AR. R. T., & Asmara, A. Y. (2017). *Creative Industry in Supporting Economy Growth in Indonesia: Perspective of Regional Innovation System.* The 3rd International Conference of Planning in the Era of Uncertainty. IOP Conference Series: Earth and Environmental Science 70.
INSEAD, & WIPO. (2017). *The Global Innovation Index 2017: Innovation Feeding the World* (10th ed.). Cornell University, INSEAD and the World Intellectual Property Organization (WIPO).

Isard, W. (1956). *Location and Space-Economy*. MIT Press.
Jacobs, J. (1972). *Economy of Cities*. Penguin Books.
Karvelyte, K. (2018). Shifting Meanings in Changing Contexts: The Role of the Creative City in Shanghai, Hong Kong and Taipei. *International Journal of Cultural Policy*. https://doi.org/10.1080/10286632.2018.1479748
Kern, P., Smits, Y., & Wang, D. (2011, January). *Mapping the Cultural and Creative Sectors in the EU and China*. A working paper in support to the development of an EU-China Cultural and Creative Industries' (CCIs) platform. KEA European Affairs.
Kong, L. (2012, June 2). *Singapore & the Making of a "Creative City"*. Chinese Creative Industries Forum 2012.
Landry, C. (2005). Lineages of the Creative City. In S. Franke & E. Verhhage (Eds.), *Creativity and the City: How the Creative Economy Is Changing the City*. NAI Publishers.
Landry, C. (2008). *The Creative City: A Toolkit for Urban Innovators* (2nd ed.). Comedia Earthscan.
Landry, C., & Bianchini, F. (1998). *The Creative City*. Demos.
Lin, C. Y. (2016). Local Planning Practices and Institutional Innovation of the Creative City: The Case of Taipei City. In J. Wang, T. Oakes, & Y. Yang (Eds.), *Making Cultural Cities in Asia: Mobility, Assemblage, and the Politics of Aspirational Urbanism*. Routledge.
Lin, C. Y., & Hsing, W. C. (2009). Culture-led Urban Regeneration and Community Mobilisation: The Case of Taipei Bao-an Temple Area, Taiwan. *Urban Studies, 46*(7), 1317–1342.
Lorenzen, M., & Frederiksen, L. (2008). Why Do Cultural Industries Cluster? Localisation, Urbanisation, Products and Projects. In P. Cooke & L. Lazzeretti (Eds.), *Creative Cities, Cultural Clusters and Local Economic Development* (pp. 155–179). Edward Elgar Publishing Limited.
Marshall, A. (1966). *Principle of Economics. An Introductory Volume* (8th ed.). Macmillan.
MITA (Ministry of Information and The Arts). (2002). *Green Paper on Investing in Singapore's Cultural Capital: A New Agenda for a Creative and Connected Nation*. Ministry of Information and the Arts (MITA). Retrieved from: http://app.mti.gov.sg/data/pages/507/doc/ERC_SVS_CRE_annex1.1(a).pdf. Accessed 12 Sep 2019.
O'Connor, J. (2010). *The Cultural and Creative Industries: A Literature Review* (2nd ed.). Newcastle upon Tyne.
Oakes, T., & Wang, J. (2016). Introduction. In J. Wang, T. Oakes, & Y. Yang (Eds.), *Making Cultural Cities in Asia: Mobility, Assemblage, and the Politics of Aspirational Urbanism*. Routledge.
Oliveira, A., & Paulino, F. (2017). European Creativity and Urban Regeneration. *The Journal of Public Space, 2*(2), 127–140.

Ooi, C. S. (2008, January). *Credibility of a Creative Image: The Singaporean Approach* (Creative Encounters Working Paper #7).

Peck, J. (2002). Political Economies of Scale: Fast Policy, Interscalar Relations, and Neoliberal Workfare. *Economic Geography, 78*(3), 331–360.

Peck, J. (2012). Recreative City: Amsterdam, Vehicular Ideas, and the Adaptive Spaces of Creativity Policy. *International Journal of Urban and Regional Research, 36*(3), 462–485.

Pratt, A. (2008). Creative Cities: The Cultural Industries and the Creative Class. *Geografiska Annaler: Series B – Human Geography, 90*(2), 107–117.

Pratt, A. C. (2009). Cultural Economy. In R. Kitchen & N. Thrift (Eds.), *International Encyclopedia of Human Geography* (Vol. 2, pp. 407–410). Elsevier.

Sasaki, M. (2007). Towards an Urban Cultural Mode of Production: A Case Study of Kanazawa, Japan. In M. Nadarajah & A. T. Yamamoto (Eds.), *Urban Crisis: Culture and the Sustainability of Cities* (pp. 156–174). United Nations University Press.

Sasaki, M. (2010). Urban Regeneration Through Cultural Creativity and Social Inclusion: Rethinking Creative City Theory Through a Japanese Case Study. *Cities, 27*(2010), S3–S9. Elsevier Ltd.

Sawangchot, V. (2016, January–June). Creative City and the Sustainable Life: A Study on the Making of Cultural Spaces in Osaka and Bandung. *Journal of Urban Culture Research, 12.* Chulalongkorn University, Tailand & Osaka City University, Japan.

Scott, A. J. (2014). Beyond the Creative City: Cognitive-Cultural Capitalism and the New Urbanism. *Regional Studies, 48*(4), 565–578.

Smith, R., & Warfield, K. (2008). The Creative City: A Matter of Values in Creative Cities, Cultural Clusters and Local Economic Development. In P. Cooke & L. Lazzeretti (Eds.), *Creative Cities, Cultural Clusters and Local Economic Development* (pp. 287–312). Edward Elgar Publishing Limited.

Statistics Indonesia (BPS). (2013). *Statistical Yearbook of Indonesia 2013.* Statistics Indonesia (BPS – Badan Pusat Statistik).

Throsby, D. (2008). The Concentric Circles Model of the Cultural Industries. *Cultural Trends, 17*(3), 147–164.

Todaro, M. P., & Smith, S. C. (2011). *Economic Development* (11th ed.). Pearson Education.

UNCTAD (United Nations Conference on Trade and Development). (2008). *Creative Economy Report.* United Nations: United Nations Conference on Trade and Development (UNCTAD).

UNCTAD (United Nations Conference on Trade and Development). (2017). *Strengthening the Creative Industries for Development in Republic of Korea.* United Nations Conference on Trade and Development (UNCTAD).

UNESCO. (2005). *International Flows of Selected Cultural Goods and Services 1994–2003*. UNESCO Institute for Statistics.
UNESCO Creative Cities Network. (2019, October 30). *UNESCO Designates 66 New Creative Cities*. Retrieved from: https://en.unesco.org/creative-cities/events/unesco-designates-66-new-creative-cities. Accessed 21 Dec 2019.
Watanabe, S. J. (2012, July 15–18). *The Historical Analysis of the 'Kunitachi Machizukuri Movement': Its Nature and the Role of Professor Shiro Masuda*. The 15th International Planning History Society Conference, Saõ Paulo, Brazil.

CHAPTER 3

The Creative City Concept: A Critique

Abstract While the creative city concept is popular and widely adopted in the last few decades, it is not without critiques. This chapter will identify some of the critiques and discuss how they will implicate the creative city-making agenda. Despite the critiques, the final section of this chapter also highlighted the need to hypothesise and test the creative city concept for Malaysian cities amid the hype within the ASEAN region.

Keywords Creative city • Critiques • Urban inequality • Malaysia • ASEAN region

3.1 Fuzzy Conceptualisation of Cultural Industries

Extending from Chap. 2, arguably, the issue of defining and delineating cultural and creative industries (which are integral towards the making of a creative city) has to be reckoned with first in any nation or city. If we cannot get the concept right at the outset, it is problematic to formulate good public policies or strategies to mobilise the cultural industries (Galloway & Dunlop, 2007) which underscore a creative city. For the past decades, debate on this has questioned the lackadaisical and loose use of 'cultural industries' and 'creative industries' which are oftentimes used interchangeably; however, they are in fact conceptually different yet little

S. L. Khoo, N. S. F. Chang, *Creative City as an Urban Development Strategy*, https://doi.org/10.1007/978-981-16-1291-6_3

attempts are made to clarify and differentiate them (Galloway & Dunlop, 2007). As the root cause, scholars acknowledged the weakness in defining and understanding 'culture' and 'creativity' which obfuscates the entire process. The definition for the term of 'culture' evolved from representing an individual's intellectual development to its application to describe the belief system, customs, expressions, etc. of people or society. It was later used to depict a set of attitudes, expressions and customs commonly shared by groups within societies (Galloway & Dunlop, 2007). When relating culture to the economy, O'Connor (2000: 5) defines cultural industries as activities that deal primarily with symbolic goods that carry primary economic value derived from their cultural value. His definition covers the 'classical' cultural industries, 'traditional arts' and art activities that are eligible for public funding. Throsby (2001), Flew (2005), Hesmondhalgh (2002) and others later look into industrial-scale production of cultural content or cultural goods within the wider creative industries agenda.

Even without fully comprehending the notions of 'culture, creativity, cultural industries and creative industries', the change in their government's disposition and strategic direction in the UK and Scotland have conveniently relabelled 'culture' as part of 'creative industries' where arts and culture are now subsumed under the creative industries agenda. This move has risked undermining the role of arts and culture. The broader concern is the indiscriminate link and fusion of the 'knowledge economy-based concept of creative industries'. Subsuming cultural industries beneath the wider knowledge economy is unwise given that the latter is void of any particular cultural content and disregards the unique characteristics of both cultural products and cultural creativity (Galloway & Dunlop, 2007: 17).

3.2 Why It Is Problematic to Quantify Culture and Creativity?

Another critique is the caution levelled against quantification of culture. With the definitional clutter and terminology confusion surrounding 'cultural' and 'creative' still unresolved, there is resistance towards blindly quantifying culture and creative industries. Scholars contended that both policy-makers and academics need to clarify this confusion which will further implicate aspects like statistical measurement and reporting (Galloway

& Dunlop, 2007) which certainly underpins the making of a creative city. Additionally, scholars also argued that the definitional turmoil to brand a creative city is oftentimes linked to the emphasis on statistical and mapping documents by city authorities (O'Connor, 2000). Such quantification of cultural and creative industries is problematic due to i) inconsistent statistical indicators across cities/regions/nations, ii) measurement of employment statistics according to outdated industrial and occupational categories, and iii) the fact that 'employment, jobs, skills and careers' have been restructured and changed radically in the past decades have impacted first and second jobs, non-paid jobs, part-time jobs and freelance work in the cultural labour force (O'Connor, 2000). Additionally, the arrival of new modes of employment and occupations with the arrival of the digital and gig economies have given new meaning to the creative and cultural industries. In the Malaysian context, the blurry conceptualisation and methodological ambiguities for mapping and measuring creative industries/creative cities were acknowledged in the academic realm (Khoo et al., 2015) but the extent upon which the plights and recommendations have been understood and permeated into the institutional framework is questionable.

3.3 Threat of Cultural Homogenisation: A Challenge to Cultural Policy

Based on diverse cultural settings, creativity is valued based on its application and focus on individual preferred creative processes. Accordingly, culture provides the fundamental materials and inputs to simplify, facilitate or modify the processes of creativity which may lead to different preferences of creative application and processes across cultures (Shao et al., 2019). While Eastern communities tend to stress on the usefulness of creativity, Western counterparts value both the usefulness as well as the novelty of creativity. The diversity of culture, where creativity is rooted, reflects the variation in creative measures and different approaches in assessing creativity (Cropley, 2000). Some evaluate the cognitive processes involved in creativity; some look into product-based creativity or creative performances/achievements; some stress on personal characteristics or personal traits; and some measure creativity based on its impact on the environment (Feldhusen & Goh, 1995; Shao et al., 2019). These variations between different individuals or communities across cultures in different cities/nations/regions triggered debates on the adoption and viability of the 'creative city' concept in different cities. Arguably, there is doubt with the existence and promise of a 'creative city' (D'Ovidio, 2016). He views 'creative city' as merely a theoretical model that looks into contemporary city by focusing on the immaterial (i.e. cultural and creative) components of the economy. The direct adoption of the concept, without considering the local idiosyncratic 'cultural DNA' (Ghilardi, 2009), might risk sacrificing local needs and values (Ljubljana Institute for Spatial Policies, 2011).

The global emphasis on cultural industries, as part and parcel of creative economy, has also challenged the fundamentals of local cultural policy. Nowadays, culture may be over-commercialised or politicised in certain cities. Advanced information and communications technology (ICT) has caused challenges to national governments in regulating global media and communication flows. Globally, governments are now engaged in socio-cultural boundary maintenance in order to deal with the negative impacts of cultural globalisation. For example, the European Community and Canada have petitioned the World Trade Organization to develop a 'cultural exception' to the disciplines of the General Agreement on Trade in Services (GATS). The objective is to defend the distinctiveness of culture and preserve global cultural diversity (Flew, 2005). In fact, pessimistic views of the relationship between culture and the economy had long existed. It is argued that culture and economy are mutually hostile. When the two are made to converge, the integrity of the former always suffers (UNDP and UNESCO, 2013). While global cities are actively promoting the creative economy, the threat of global cultural homogenisation is always a concern. The extent upon which aspiring Malaysian creative cities are confronted with the challenges of global cultural homogenisation will be elucidated in Chaps. 5, 6, 7 and 8.

3.4 Widening Urban Economic Disparity and Socio-Spatial Inequality

When dissecting the creative city concept within the discourse of social inclusion, liveability and urban sustainability, concerns pertaining inequality and disparity loom to the forefront (Pratt, 2011). In Malaysia, cities are neither standalone nor isolated but they are increasingly agglomerating in clusters as evidenced through national agglomeration and corridor initiatives (i.e. North Corridor Economic Region,

Iskandar Region). The regionalist functions of Malaysian city regions, particularly major ones like Kuala Lumpur Conurbation, George Town Conurbation and Johor Bahru's Iskandar Region, have unleashed the potentials of these city regions as sites of economic competitiveness and social reproduction. Viewing Malaysian cities in this manner is not new. Predecessor cities in North America and the UK (i.e. Cambridge) have undergone these processes but with a quandary of mixed outcomes. A case in point is Austin when it rose to its heydays from the mid-1990s to the start of the twenty-first century. Though economically vibrant and amenity-rich (endorsing the checklist qualities of Florida's utopian creative city), underneath those glories were pressing urban issues like escalating housing prices, bifurcated urban social structure, gentrification and environmental degradation. Such urban realities and trepidations challenged policy responses pertaining the real meaning of the 'good life' (McCann, 2007: 190). Some creative initiatives are said to be simply a disguised real estate revaluation that causes displacement and undermines more spatially decentralised cultural vitality (Kunzmann, 2004). Critics also argue that creative strategies have led to undesirable consequences due to the production of entertainment or tourism-oriented commercialised urban spaces, which may eventually promote greater social stratification (Zukin, 1995). The downsides of creative cities are oftentimes associated to the econo-centric or entrepreneurialism approach adopted by cities where neo-liberalist reforms tend to prioritise, commodify and escalate the price and value of cultural endowments thus rendering urban areas as sites of wealth accumulation and competition between the rich and poor. Critics, especially those in the scholarly realms (Pratt, 2011; Kunzmann, 2004), have cautioned that if implemented carelessly by urban managers, the creative city concept will be deglorified when economic and socio-spatial inequalities heighten.

3.5 An Urban-Centric, Urban Biased and Exclusive Concept

The creative city route is popular yet equally contentious in terms of advocating inclusive and holistic development across urban places and spaces as well as between urban and rural settings. The concept itself leans towards urban centricity which in turn compounds urban biasness. Globally, the urban milieu is experiencing an increase in terms of population, investment and ancillary activities, thus, making it a much sought-after site for power and capital. Many countries in the developed world have strategised their cities based on urban neoliberalism which inevitably create urban biasness which further deepens fissures between urban and rural areas. The creative cities concept if not fully understood by political leaders and urban manager could then pose a challenge to its implementation and outcome. In Malaysia, noticeably, socio-spatial disparity between East and West Malaysia has existed since political independence (1957) and continues to persist. Evidently, the East and West coasts of Peninsular Malaysia are socio-spatially different as the West coast is more advanced than the East coast. Income levels are significantly higher in cities on the West coast of Peninsula like Kuala Lumpur, George Town, Johor Bahru compared to cities like Kota Bahru, Kuala Terengganu and Kuantan on the East coast.

Western scholars have debated the adjective 'creative' that precedes the word 'city' and the way it has been unsettling for many because such a prefix suggests and implies that creativity and all its manifestations are purely urban-centric. However, creativity prevails in all realms whether in urban or rural settings. With emerging scenarios of territorial development where urban and rural areas are opposites no more, the immediate line of enquiry would be the validity of a creative city strategy. Another logical enquiry is to question and challenge Florida's spiky world theory stating that the 'creative class' is purportedly agglomerating in cities due to creative energy and vibrancy in cities. Does this imply that non-urban, peripheral and rural places are deemed to be 'less creative'? At the occupational level, further critiques have challenged the definition and classification of Florida's creative class by questioning whether creativity is absent in other occupations not defined and acknowledged within the creative class taxonomy. In fact, scholars highlighted that it is the urban poor who are the 'real creative class' given that they frequently need to practise reflexivity and creativity to brace life's struggles (Wilson & Keil, 2008: 842). Arguably, creative city as an urban development strategy can cause urban exclusion due to variances in political ideology and distinctive socio-spatial segregation (Pratt, 2011).

3.6 The Creative Class: A Controversial Terminology

In the literature, the term 'creative class' coined by Florida (2002, 2004) is greeted with ambivalence where it is both celebrated and abhorred. While the creative city concept is advocated to advance the noble tenets of urban inclusion through UNESCO's Creative Cities Network and cultural diversity in the Universal Declaration of Cultural Diversity 2001, the concept is also flawed due to Florida's obfuscating creative class definition that divides and delineates across class, colour and creed. Arguably, labelling of creative class delineated by class, culture and creativity is potentially damaging and causes class divide (Pratt, 2011).

In 2013, UNDP and UNESCO highlighted the decline of creative class paradigm across cities in the United States, Northern Europe and East Asia. Florida's thesis was criticised due to insufficient supporting empirical evidence and the lack of guidance to facilitate necessary conditions for skilled and creative talents to congregate and stay in a place to become key agents in local and regional development. At some point, even Florida (2013) himself had to admit the limits of the creative class when seeing how the clustering of talents failed to generate trickle-down benefits but mainly benefited disproportionately the higher-skilled knowledge professionals and creative workers who earned higher wages. Florida (2017) concluded that clustering of talent and economic assets generate lopsided, unequal urbanism, in which only benefit a handful of superstar cities and elite neighbourhoods while many other places are still stagnated or fall behind. Besides Florida, there are sceptics who are doubtful whether cities could continuously attract the 'creative class' to stay for long (Ljubljana Institute for Spatial Policies, 2011). Apart from the above controversies, scholars also criticised Florida's over-simplistic idea that the moment a creative class converged in a particular location, it will activate the cultural energies and innate entrepreneurialism that will then stimulate local economic development. As argued by Scott (2006: 11), Florida's argument failed to consider the 'complex synchronic and diachronic interrelationships' that need to be available prior to the emergence of a vibrant creative environ. Scott (2006) further elucidated that the bare minimum to construct a creative city will require the formation of a local production system, upskilling/attracting requisite manpower, suitable programming of urban space and to ensure all the elements involved are working harmoniously together.

3.7 Testing the Creative City Concept for Malaysia

Albeit the above critiques, a creative city has to live up to its name as a city that possesses unique culture, creativity and innovation to propose urban solutions when confronted with new urban issues/challenges (Landry, 2000). Despite the litany of criticisms, especially within the academic realm, the creative city concept is still widely adopted globally. Hence, this is the time to test the concept for Malaysia. While the creative city strategy has garnered much popularity in the Western world (i.e. Europe, North America) and more recently in East Asia (i.e. South Korea, Japan, China), cities in Southeast Asia (SEA) are lagging behind in the race. Insofar, only a handful of SEA cities have hopped on the creative city bandwagon with Thailand having four creative cities to date and Singapore being bestowed as Creative City of Design in 2015. Malaysia which is sandwiched in the middle does not fall short in possessing unique cultural and creative endowments to go along the creative city route, but to date Malaysia has not secured a spot as a designated UNESCO Creative City. The fragmented efforts and lack of strategic direction to steer the emerging creative economy in cities have placed Malaysian cities at a crossroads. Arguably, Malaysian cities could collaborate with Singaporean and Thai cities by identifying and showcasing their creative and cultural endowments collectively to unleash and reap a regional spillover effect. To this end, the Southeast Asian Creative Cities Network which is an alliance informally established in 2014 should assume a bigger role to coordinate and manage collaboration across Southeast Asian cities. The local creative and cultural assets, native skills/knowledge and indigenous cultural creativity of different SEA countries could be harnessed to showcase the hidden gems of SEA in achieving urban and regional sustainability. Broadly, this is aligned with the principle of agglomeration economies that underpin the creative city agenda which is also apt to be adopted at a regional level. While the modern world system can be portrayed as a 'mosaic of complementary and competing regional economies (Veltz 1996) (cited in Scott, 2006: 13), this scenario is in fact advantageous to creative city-making where competitions of these kinds are advantageous to cities with unique creative capabilities that form their niches. Apart from competition, the global landscape has also encouraged collaboration where dynamic firms in creative cities are involved in fostering global networks of creative partnerships (i.e. strategic alliances, joint ventures, co-productions) which in turn synergises distinctive mix of skills, ideas and talents from diverse agglomerations with varied cultural beliefs and creative capacities (Scott, 2006: 13).

Transnational and cross-cultural collaboration in creative initiatives and developments could eventually expand the 'clustering' effect of creative talents within the SEA regional creative hub which are formed by multiple creative cities or suburbs under a 'Mega-Urban Region of Creative Cities' regional bloc of sorts. According to Florida, Gulden and Mellander (Florida et al., 2008), mega-regions are integrated sets of cities and their surrounding suburban hinterland where labour and capital can be reallocated at very low cost. Instead of nation-states, it is now cities and also mega-urban regions that are functioning as central engines of economic growth with agglomeration of talents, productive capability, innovation and market (Jacobs, 1969, 1984; Florida et al., 2008). Linking these propositions to Scott's (2006) arguments, collaborations and shared visions among SEA creative cities can be leveraged to be the key catalyst for collective competitiveness and prosperity of the SEA region. Besides sharing the concentration of population and economic outputs, despite their national borders, mega regions are likely to develop similar social mores, cultural tastes and even political leanings (Florida et al., 2008). Thus, working together among SEA cities in establishing a (regional) 'creative ecology' with a well-developed (extended) 'creative platform', supported by the hard and soft creative infrastructure (within the region), plays a vital role to generate a flow of ideas and inventions among the (regional) communities and to attract the clustering of 'creative class'. Consequently, this could lead to the rise of 'Creative Mega-Urban Region of South East Asia' – a regional bloc that shares and harnesses their similar creative significance and characteristics. This could happen through collaborations in creative city research, practices/initiatives as well as funding and support among counterparts within the region. For instance, Singapore being bestowed a Creative City of Design status in 2015 has some influential traction on cities within the region. Malaysia, being the closest neighbour will be the first to feel the effect by questioning Singapore's position. The pertinent question of Singapore's position whether as competitor or collaborator within the South-East Asian region would be worthy of exploration. This would be teased out in Chap. 8 when exploring the scenario for Johor Bahru which is Singapore's closest neighbour.

Success stories and collaborative opportunities from neighbouring SEA countries like Singapore, Thailand, Indonesia, Vietnam and such may become the motivation to inspire and encourage Malaysian creative practitioners, stakeholders and investors in Malaysian cities to explore further the feasibility and potential of local creative assets. The readiness of the local production system and the institutional framework to mobilise this endeavour are integral aspects to understand the dynamics and state-market nexus towards inclusive and sustainable creative city-making. The developmental milestones experienced by the competitors (or collaborators) within the same SEA region with rather similar creative and cultural backgrounds and significances could serve as relevant references and guidance. However, unlike Singapore, Brunei or other South-East Asian cities that are more homogenous in their population and demographics composition, Malaysia is unique due to the nation's rich cultural diversity. The three main ethnic groups of Malay, Chinese and Indian coupled with other indigenous groups (i.e. *Dusun, Kadazan, Bajau*) offer Malaysian cities a wide and diverse spectrum of cultural and creative assets that not only create local employment and generate revenue if harnessed carefully, but also unearth and unravel the identity, pride and unique character of Malaysian cities that are location-specific, place-bound and not found elsewhere. The following chapters endeavour to comprehend the scenario for four aspiring Malaysian creative cities (i.e. Kuala Lumpur, George Town, Ipoh, Johor Bahru) where each city will be discussed separately in Chaps. 5, 6, 7 and 8 respectively. These cities are selected based on their city's strategic direction and the vibrancy of creative endeavours present in each city. Before illustrating the case of each individual city, the next chapter will present an overview of Malaysia's developmental stages, related national policies and challenges that will influence and shape the nation's creative city agenda.

References

Cropley, A. J. (2000). Defining and Measuring Creativity: Are Creativity Tests Worth Using? *Roeper Review, 23*, 72–79. https://doi.org/10.1080/02783190009554069

D'Ovidio, M. (2016). The Creative City Does Not Exist. In *Critical Essays on the Creative and Cultural Economy of Cities*. Ledizioni Ledi Publishing.

Feldhusen, J. F., & Goh, B. E. (1995). Assessing and Accessing Creativity: An Integrative Review of Theory, Research, and Development. *Creativity Research Journal, 8*, 231–247. https://doi.org/10.1207/s15326934crj0803_3

Flew, T. (2005). Creative Economy. In J. Hartley (Ed.), *Creative Industries* (pp. 344–360). Blackwell Publishing Ltd.

Florida, R. (2002). *The Rise of the Creative Class: And How It's Transforming Work, Leisure, Community and Everyday Life*. Basic Books.

Florida, R. (2004). *The Rise of the Creative Class*. Basic Books.

Florida, R. (2013, January 30). *More Losers Than Winners in America's New Economic Geography*. Bloomberg CityLab. Retrieved from: https://www.bloomberg.com/news/articles/2013-01-30/more-losers-than-winners-in-america-s-new-economic-geography. Accessed 30 Oct 2019.

Florida, R. (2017, April 9). *The Roots of the New Urban Crisis*. Bloomberg CityLab. Retrieved from: https://www.bloomberg.com/news/articles/2017-04-09/richard-florida-and-the-new-urban-crisis. Accessed 30 Oct 2019.

Florida, R., Gulden, T., & Mellander, C. (2008). The Rise of the Mega-Region. *Cambridge Journal of Regions, Economy and Society, 1*(3), 459–476.

Galloway, S., & Dunlop, S. (2007). A Critique of Definitions of the Cultural and Creative Industries in Public Policy. *International Journal of Cultural Policy, 13*(1), 17–31.

Ghilardi, L. (2009). Nature or Nurture? Rethinking the Creative City. *ERA, 21*(4).

Hesmondhalgh, D. (2002). *The Cultural Industries*. Sage.

Jacobs, J. (1969). *The Economy of Cities*. Random House.

Jacobs, J. (1984). *Cities and the Wealth of Nations: Principles of Economic Life*. Random House.

Khoo, S. L., Badarulzaman, N., Samat, N., & Dawood, S. R. S. (2015). Creative Cities Research in Penang, Malaysia: A Review of Conceptual and Methodological Framework. *Planning Malaysia, XIII*, 1–18.

Kunzmann, K. R. (2004). Culture, Creativity and Spatial Planning. *The Town Planning Review, 75*(4), 383–404. Liverpool University Press.

Landry, C. (2000). *The Creative City: A Toolkit for Urban Innovators*. Earthscan.

Ljubljana Institute for Spatial Policies. (2011, November). *Potentials of Creative Urban Regeneration. Final Study*. IPoP, Institute for Spatial Policies. Retrieved from: http://creativecitiesproject.comune.genova.it/en/output/2013-05/02/14-Ljubljana/Potentials_of_creative_urban_regeneration_Ljubljana_EN.pdf. Accessed 31 Oct 2019.

McCann, E. J. (2007). Inequality and Politics in the Creative City-Region: Questions of Livability and State Strategy. *International Journal of Urban and Regional Research, 31*(1), 188–196.

O'Connor, J. (2000). The Definition of the 'Cultural Industries'. *The European Journal of Arts Education, 2*(3), 15–27. Retrieved from: https://eprints.qut.edu.au/43877/2/43877.pdf. Accessed 30 Oct 2019.

Pratt, A. (2011). The Cultural Contradictions of the Creative City. *City, Culture and Society, 2*(3). ISSN 1877-9166.

Scott, A. J. (2006). Creative Cities: Conceptual Issues and Policy Questions. *Journal of Urban Affairs, 28*(1), 1–17.

Shao, Y., Zhang, C., Zhou, J., Gu, T., & Yuan, Y. (2019, May). How Does Culture Shape Creativity? A Mini-Review. *Frontiers in Psychology, 10*, Article 1219.

Throsby, D. (2001). *Economics and Culture*. Cambridge University Press.

UNDP, & UNESCO. (2013). *United Nations Creative Economy Report 2013: Widening Local Development Pathways*. UNDP.

Wilson, D., & Keil, R. (2008). The Real Creative Class. *Social & Cultural Geography, 9*(8), 841–847.

Zukin, S. (1995). *The Cultures of Cities*. Blackwell.

CHAPTER 4

The Malaysian Scenario: Setting the Context

Abstract Before we delve into the viability of creative city as an urban development strategy for Malaysian cities, this chapter will provide an overview of the Malaysian context where macrolevel social, economic and political structures and national policies will be discussed. This overview serves to position the Malaysian context pertaining to creative city-making. The chapter will also discuss the conundrums surrounding the concept, classification and taxonomy related to Malaysia's creative industries and creative economy. The challenges encountered in creative city-making will be fleshed out too. The chapter concludes by questioning and deliberating the efficacy and sustainability of the creative city strategy for Malaysian cities.

Keywords Malaysia • Creative city • Macro policies • Challenges • Urban sustainability

4.1 Malaysia: An Overview

Malaysia comprises of Peninsular Malaysia and East Malaysia. The nation is among the most vibrant developing countries in Southeast Asia. With a population size estimated at 32.6 million in 2019, the Malaysian citizenry consists of 69.3% *Bumiputera*,[1] 22.8% Chinese, 6.9% Indians and 1% others (DoSM, 2019). With an aspiration to achieve developed country

[1] Term used in Malaysia to describe Malays and *Orang Asli* (indigenous people of Malaysia).

S. L. Khoo, N. S. F. Chang, *Creative City as an Urban Development Strategy*, https://doi.org/10.1007/978-981-16-1291-6_4

status, Malaysia is expected to reach the high-income threshold between 2020 and 2024 (World Bank, 2017). As elucidated briefly in Chap. 1, Malaysia also followed the conventional path of relying on agriculture and natural resources in the early years of nation-building after political independence in 1957.

Since 1950, Malaysia launched the nation's five-year development plan to drive national development. The development plans are developed to look at socio-economic perspectives with implementation of strategic policies at every five-year milestone (Lee & Lee, 2017). The period right after political independence in 1957 was directed towards nation-building, infrastructural development and development of rural population, trade and agricultural-based industries (Siddiqui, 2012; Lee & Lee, 2017). Subsequently in the 1960s, the Malaysian government's focus was on employment generation, development of socio-economic components as well as economic diversification where there were attempts to diversify the economy towards higher value manufactured goods beyond tin and rubber. Following the 1969 racial riot, the government launched the New Economic Policy in 1971 with the aim to restructure the Malaysian society and eradicate poverty. Following suit, the government ventured into the oil industry during the 1970s. Gradually, in the 1980s, the Malaysian government under then Prime Minister, Mahathir Mohamed, started to implement neoliberal reforms where state-owned enterprises were restructured and privatisation became fashionable. The receding role of the Malaysian state became obvious across all sectors. During this period too, the political leanings of the state was less towards the West but to the East thus emphasis on the 'Look East' policy to countries like Japan. Later, under the Sixth Malaysia Plan (1991–1995), the government announced the Vision 2020 and launched the National Development Policy (1991–2000). It was during this period too that the Malaysian government started to look into sustainable and balanced development with a vision to transform Malaysia into a developed nation by 2020. In 2001, the National Vision Policy was launched under the Eighth Malaysia Plan (2001–2005). Among the key elements during this period is the shift of Malaysia's focus towards developing a knowledge-based economy where people's knowledge is viewed as the new economic driver. This is a defining moment when positioned within the creative economy discourse where scholars (Galloway & Dunlop, 2007) have cautioned about the rise of the knowledge economy and how cultural industries are sometimes conveniently and mistakably subsumed beneath it. Emulating the routes of many advanced societies where 'cultural industries' gradually got intertwined with the 'creative industries' and 'knowledge economy' definitional conundrum, Malaysian policy-makers are also obfuscated by the definitions and delineations of Malaysia's cultural and creative industries. This has a bearing towards the development of Malaysia's cultural and creative industries where any misguided policy direction for these two diverse industries will underscore the formation of Malaysian creative cities.

Subsequently, Malaysia's Ninth Malaysia Plan (2006–2010) still carried on the notions of a knowledge-based economy with emphasis on human capital development to address issues of socio-economic inequities. Later, the Tenth Malaysia Plan (2011–2015) witnessed many transformative processes to drive national development led by the New Economic Model. The current Eleventh Malaysia Plan (2016–2020) has a more holistic and inclusive proposition where the Malaysian government is now looking into post-industrialisation or deindustrialisation strategy. Noting that industrialisation and modernisation can cause backlash, the attention now is towards green and sustainable development that prioritises inclusivity, well-being and growth of the people instead of merely GDP growth (Economic Planning Unit, 2015).

To achieve the high-income threshold to be on par with other developed nations, the Malaysians government has initiated efforts to develop the national creative and innovative economy by shifting from investment-led to innovation-led growth. The government has diverted focus towards productivity enhancements and new value creation instead of capital accumulation and technology adoption (Lee & Lee, 2017; KRI, 2018). The investment in knowledge, innovation and creativity has become the new economic catalyst towards building creative competitiveness among Malaysians. This shift in Malaysia's strategic direction to embrace the knowledge economy is similar to the situation in the UK, Australia and Europe during the 1990s. Arguably, in the Malaysian case as well, the cultural sector has somewhat been subsumed within the wider creative industries and knowledge economy agenda rendering culture as another 'knowledge economy asset'. Before the government could adequately harness the economic potentials of Malaysian arts, the language of 'creativity' and 'knowledge economy' surfaced incessantly in Malaysian policy rhetoric post-2000. This backdrop will influence and shape the agenda of creative city-making in Malaysia.

4.2 Grounding Creative City Narrative in Malaysia

In Malaysia, grounding the creative city narrative is a challenge by itself. Just like tackling other gargantuan concepts like affordable housing, smart city and such, the notion of creative city is similarly huge, ambiguous and rather new in Malaysia in terms of attaching a formalised label to the field and the activities within it. Yet it cannot be dismissed that a myriad of diverse stakeholders (i.e. policy-makers, urban managers, cultural workers, communities) in Malaysia's pluralistic society has in one way or another engaged themselves in creative and cultural activities. They are in fact creative and cultural workers in their own right. A more accurate depiction of current state of affairs would be the shift in focus of Malaysia's overall strategic direction with the emphasis on knowledge-based, creative and cultural-led urban development strategies. These global trends are being emulated by some Malaysian cities with much fanfare as they transform to become creative cities. Notwithstanding, Malaysia is also caught in the similar situation of finding it problematic to delineate and decipher individually what 'creativity' and 'culture' means – a dilemma compounded (or perhaps enriched) by Malaysia's multi-ethnic backgrounds and cultural diversity. This then poses the question of *"What culture? Whose culture?"* and the extent upon which creativity is present, harnessed and further cultivated in these diverse cultures (Jenkins, 2019).

In the attempt to contextualise the creative city agenda in Malaysia, it is pivotal to first comprehend some key national policies conceived after Malaysia's independence in 1957. As a newly decolonialised nation state eager to find its own identity, Malaysia formulated the National Cultural Policy in 1971. The formulation of the policy considered the historical background of the region, the function of the country as a meeting point for several major civilisations and its role as an important trading centre for 2000 years. Regionally, Malaysia together with Indonesia, the Philippines, Singapore, Brunei, Thailand, Kampuchea, South Pacific Islands and Malagasi were key centres of Malay Civilisation or culture. Collectively, this region long played an integral role to cultivate and develop the Malay culture. The culture of the region's indigenous people is deciphered by commonalities in the Malay-Austronesian language family, history, geographies, natural resources, arts and values. In Malaysia, it started in Malacca by the great Malacca Sultanate and the identification and use of Malay language as lingua franca. By identifying the Malay culture as the way of life as well as the identity and the yardstick of individuality of Malaysia's indigenous population, thus, the Malay culture became the crux of Malaysia's National Cultural Policy (Nordin & Hussin, 2011: 303).

4.2.1 National Cultural Policy (NCP)

Just like in most countries, Malaysia's cultural and creative industries are intrinsically linked to the nation's National Cultural Policy. For a culturally diverse and pluralistic nation like Malaysia, NCP is difficult to fathom and often becomes contentious (Mandal, 2008). The government initiated the policy as a measure to encourage national unity among Malaysia's multiracial society towards strengthening a Malaysian identity. The formulation of the NCP was based on three principles. First, the culture of the indigenous group, namely the Malay culture, formed the basis of the policy. Second, the policy considered the multi-cultural aspects of Malaysians by including suitable and acceptable cultural elements of the Chinese, Indians and others. Lastly, as the official religion in Malaysia, Islam is the important component of NCP. The policy courted contentious debates throughout the years of its implementation. Malaysian scholars stood on extreme ends of the continuum with some advocating that the policy continues to advocate Malay identity as the basis of national identity (Aziz, 1987; Ismail, 1990) while others argued that it should reflect a Malaysian way of life amid a multi-ethnic and pluralistic society (Rowland, 2004; Zawawi, 2004). The historical openness of indigenous Malay culture to external influences and acceptance of multicultural elements of different non-indigenous groups were amongst the key factors that led to the implementation of NCP (Ismail, 1990; Zawawi, 2000; Rowland, 2004). These debates are not only about citizenship and the rights of groups but are related to the sense of belonging among national communities (Mandal, 2008).

Apart from the Malay culture, interestingly, it is through Malaysia's historic role as a trading centre during the Malacca Sultanate that initiated the introduction, interaction and assimilation of diverse cultures from all over the world, both East and West. The presence of these foreign cultures gradually assimilated and shaped Malaysia's cultural eclecticism we experience today. This is due to the fluidity of culture as a dynamic and ever-changing process of on-going adaptation and acculturation – a principle that corresponds to the real-life context of the pluralistic population and cultural multiplicity of Malaysian culture. While Chinese, Indian, Arabic and Western cultural elements are duly considered in NCP, their acceptance will depend on whether they conflict with the basic principles of the Federal Constitution, respective State interests and the official religion, Islam (Nordin & Hussin, 2011: 303). With known historical evidence that Islamic values were supported and embraced by a huge portion of the populace in the region and Islam being Malaysia's official religion, Islam was the main aspect in the establishment of a national culture (Nordin & Hussin, 2011: 304).

This overall scenario influences Malaysia's art and cultural sector. As succinctly highlighted by Rowland (2004), the centrality of Malay culture has consequently shaped the direction of arts and cultural funding initiatives by the government. In the 1980s, Malay language cultural activities were being prioritised with more frequent access to government support through funding and promotional activities such as festivals, awards and competitions. Nevertheless, since 1991, the new *Bangsa Malaysia* was introduced and advocated by then Prime Minister, Dr. Mahathir Mohamad, towards building more competent and prosperous Malaysian nationalism that rises in cooperation through inter-ethnic alliances, globalisation and internationalisation. Indirectly, since then, NCP's centrality for Malays has been contested (Mandal, 2008). Despite the push for inclusiveness and globalisation, the NCP somehow holds the government's capacity to engage beyond discourses that were alienating to the national culture. The NCP's constraints or limitations of tolerance and openness towards new and foreign cultural elements became the significant challenge in steering the development of creative cities or creative economy in Malaysia. Coincidentally, 'tolerance' is one of Florida's 3 Ts (i.e. Technology, Talent and Tolerance) key elements of urban competitiveness and attractiveness (Florida, 2005). Nonetheless, to boost Malaysia's tourism and cultural sectors, the Ministry of Tourism, Arts and Culture (MOTAC) announced the formulation of a new NCP (based on NCP 1971) in 2019 which will focus on multi-ethnic cultural inclusion towards promoting Malaysia's national identity (The Borneo Post, 2019). Apart from Malaysia's cultural backdrop, the nation's development trajectory (as the nation globalises) also has its bearing towards shaping Malaysian creative industries and creative cities. This leads us to two other national policies, namely, the New Economic Policy (NEP) and National Urbanisation Policy 2 (NUP2) as elucidated below.

4.2.2 New Economic Policy (NEP)

In the same year (i.e. 1971) when NCP was put in place, the New Economic Policy (NEP) was officially enforced in respond to the 1969 racial riot. Similar to NCP, NEP aimed to build national unity. It focused on poverty eradication and economic restructuring towards eliminating the identification of ethnicity with economic function (Thillainathan & Cheong, 2016). Strategies under NEP included export-oriented industrialisation and interventions in the market through investments at the federal and state level. While NEP aimed to reduce poverty, the policy also caused income inequality between different ethnic groups as well as between rural and urban settings. In dealing with the 1980s economic recession, the government turned to privatisation to provide more opportunities to *Bumiputera* community to participate in privatised activities for equity ownership and management (World Bank, 1999; Pua, 2011).

No doubt, the NEP drove national economic and development planning in Malaysia. Overall, it has significantly reduced national hardcore poverty and improved income distribution. The middle class has grown. With a diversified economy, Malaysia has attained the status of an upper middle-income nation. Nevertheless, NEP remains the most debated policy in the country since its inception (Thillainathan & Cheong, 2016; Zainuddin, 2019). State intervention with quotas for equity ownership and management for *Bumiputera* has somehow distorted market signals. The equity recipients have rarely acquired the entrepreneurial skills to manage the private entities. The strong state's support in the background has undermined corporate governance, but generated the reliance on a political patronage bred system. Consequently, this might impact national vulnerability during economic crises (Thillainathan & Cheong, 2016). This led to the need to restore the private sector's role in market, thus, the rise of neo-liberalism in Malaysia during the 1980s. For Malaysia's way forward, Zainuddin (2019) urged for a new NEP with needs-based approach instead of race-based approach with consideration on the globalised developmental trend that recommends capitalisation on Malaysia's strengths and comparative advantages in our local industries, including the halal industry, agriculture, tourism, palm oil and also the arts and cultural industries.

4.2.3 National Urbanisation Policy 2 (NUP2)

Formulated and launched since 2006, the National Urbanisation Policy (NUP) (2006) is in its second edition, namely NUP2 (2016–2025). It is an important policy to drive and manage the planning and development of sustainable urbanisation processes in the country, with emphasis on balanced development for physical, environment, social and economic perspectives. This emphasis is crucial given that Malaysia's growing urban population is expected to reach 27.3 million by 2025. The figure represents almost 80% of the Malaysian population (Haji Omar, 2016). Broadly, the NUP aims 'to create a visionary city with a peaceful community and living environment through sustainable urban development'. With this aim, the policy has identified six objectives: (i) to develop a planned, quality, progressive and sustainable city; (ii) to develop and strengthen a competitive urban economy; (iii) to create a conducive environment in order to encourage social development; (iv) to eradicate urban poverty; (v) to strengthen the planning, implementation and monitoring system; and (vi) to strengthen urban management and administrative institutions. The emphasis of the NUP2 towards inclusive, resilient and sustainable urban development are in line with the six strategic thrusts specified in the Eleventh Malaysia Plan (2016–2020) as well as the Sustainable Development Goal (SDG) No. 11 and the New Urban Agenda (NUA). Thus, when investigating the applicability of creative city notions in Malaysia, it is important to ground the debate within the NUP2 and also global principles like the SDGs and NUA. Broadly, the above three policies reviewed are intrinsically linked to Malaysia's political economy and the way the nation integrates into the global economy.

Integration into the global economy inevitably implies the emergence of free economy, more foreign direct investment, global competition and the need to produce more well-rounded scholars and knowledge graduates in Malaysia to compete in the global arena (Nordin & Hussin, 2011: 370–371). Human capital is akin to people and is viewed as the new factor of production apart from capital, industry and infrastructure. Following suit, human capital development in Malaysia moved to the forefront of policy agenda and was propagated widely during Abdullah Ahmad Badawi's time as Prime Minister. It was perceived that a nation's competitive edge will be enhanced through human capital development. For a developing nation like Malaysia, development of human capital has placed much emphasis on fields related to Science and Technology such as biotechnology, nanotechnology, allied health sciences and such. Syllabi related to arts, social sciences, cultural and creative education somewhat took a backseat in Malaysia (Frost & Sullivan, MyPAA and TFCC, n.d.). Additionally, the rule of the game has changed where competition is no longer among nations but cities (Kharas et al., 2010). This notion corresponds to the creative city argument where twenty-first century does not only celebrate the importance of human capital, knowledge and innovation; but it also reinstates and reinforces the role of urban location in economic development (Florida, 2002, 2004, 2007, 2008) debunking scholars like Friedman (2006) who argued that forces of globalisation have rendered the demise of location. Based on Florida's contention, a city that can brace all perils of global competition will be appealing, attractive and 'spiky' where human talent, particularly the creative class, will want to agglomerate to create the synergistic clustering force of a creative city. As put forth by Landry (2012), a creative city is a city that can leverage upon its strengths as well as address the city's new challenges as the city morphs and transforms.

Against this myriad of diverse definitions, unquestionably, Malaysia encounters the challenge of defining the creative city concept for the Malaysian context. This is compounded by the fuzziness in deciphering Malaysian creative and cultural industries, and further complicated by the way the meaning of creative/cultural industries morphs in tandem with the developmental trajectory of the nation. Rather similar to situations in Singapore and some Western cities (Kong, 2012; Galloway & Dunlop, 2007), Malaysian creative industries experience this conundrum of having 'culture', 'cultural industries' and 'cultural creativity' deeply embedded within it. However, the subsequent shift in national strategic direction towards a knowledge-based economy has espoused the elements of creativity and innovation to be different from those alluded earlier in the cultural domains.

In Malaysia, the National Creative Industry Policy (NCIP) has links with the National Department for Culture and Arts since the 1970s. Back then, it functioned as a government framework of guidelines to ensure frequent emphasis on Malaysian arts and culture. It was a social contract that was constantly updated with input from industry players. During the Sixth Malaysia Plan in 1991, however, the policy morphed to suit the strategic directions outlined under Vision 2020.[2] Subsequently, in 2009, the Prime Minister announced the formulation of NCIP under the Ministry of Information, Communications and Culture (MICC) (MICC, 2010). Increasingly, Malaysia's cultural industries were placed at a crossroads when the overarching influences of the knowledge-based economy started to encroach into their boundaries. The situation is such where the NCIP has origins and links with the National Department for Culture and Arts (which is cultural-based), but the broader aspirations of contemporary creative industry policy are attuned to goals in a knowledge-based economy. This scenario will have policy and practical ramifications that are disadvantageous to Malaysia's cultural industries. Arguably, the Malaysian scenario during the 1990s onwards mirror the situation as disputed by Western scholars (Galloway & Dunlop, 2007) – a scenario where Malaysia was visibly embracing a 'knowledge economy-based concept of creative industries' with ambiguity and confusion in fathoming specific cultural content and distinctive characteristics of cultural creativity, processes and products for Malaysia's cultural and creative industries. This conundrum is further highlighted below when the Malaysian classification and taxonomy of the creative industries have somewhat fused 'creativity' in high-tech industries like (i) multimedia as well as traditional sectors such as (ii) cultural arts and (iii) cultural heritage. Scholars have criticised this aspect by arguing that the NCIP failed to develop and instil creativity. Instead, the policy has conveniently assumed that creativity is present in certain types of cultural goods as evident in the way the adjective 'creative' appeared in the three categories identified by the policy. With or without the adjective 'creative', the categories are still coherent implying that the prefix 'creative' has been insufficiently defined. Although attaching the prefix 'creative' to the 'cultural heritage' category has been debated as to whether preservation and restoration are in fact creative activities (Barker & Lee, 2017: 24), creative city literature in East Asian countries like Japan in particular have leveraged 'cultural creativity' as part of their cultural mode of production and consumption (Sasaki, 2004, 2010, 2011).

[2] National Creative Industry Policy (NCIP). Source: https://www.arteri.com.my/learn/policies/dikn/ (Accessed on 20 August 2020).

4.3 Classification and Taxonomy

In Malaysia, to date, the only relevant and official policy related to creative industries is the National Creative Industry Policy (NCIP) (Barker & Lee, 2017: 21). Similar to most creative industries policies worldwide, Malaysia also referred to UK's DCMS definition (as illustrated in Chap. 2) for the local context. The only difference, however, is the classification of the sub-sectors where Malaysia's National Creative Industry Policy identified 11 sub-sectors that are divided into three (3) categories: (1) Multimedia, (2) Cultural Arts, and (3) Cultural Heritage (refer Table 4.1) (MICC, 2010). More specifically, Malaysia's NCIP focuses on digital media which is also the emphasis of government agencies such as Multimedia Development Corporation (MDeC).

During the recent Creative Economy 2021 Forum held on 20 October 2020, YB Dato' Sri Mohd Uzir Mahidin, the chief statistician of the Department of Statistics, Malaysia (DoSM) reported that under the Malaysia Standard Industrial Classification 2008 (MSIC 2008), a total of 45 classified industries are grouped as creative industries (CI) based on NCIP's three main categories (i.e. multimedia, cultural arts, cultural heritage). Table 4.2 shows some key statistics about Malaysia's creative industries.

While statistics are compiled and analysed based on NCIP's classifications at one end, it is crucial to have a standard and mutual acknowledgement of definition/classification of creative industries among different government departments. Obviously, there is inconsistent comprehension between DoSM and the current Ministry of Communications and Multimedia (MCM) when the Minister of MCM classified CI differently

Table 4.1 NCIP classification of the creative industries – 11 sub-sectors

Multimedia	*Cultural arts (Seni Budaya)*	*Cultural heritage (Warisan Budaya)*
■ Film & TV ■ Advertising ■ Design ■ Animation & Digital Content	■ Craft ■ Visual arts ■ Music ■ Performing arts ■ Creative publishing ■ Fashion & textiles	■ Cultural heritage

Source: Compiled by authors from the NCIP

Table 4.2 Key statistics of creative industries (CI) in Malaysia

	No. of establishments as per September 2020	*Share of employment in 2019 ('000)*	*Contribution to GDP in 2019 (RM million)*
Multimedia CI	10,802	200.4 (23.3%)	14.91 (50.7%)
Cultural arts CI	20,898	651.8 (75.8%)	14.46 (49.2%)
Cultural heritage CI	92	7.7 (0.9%)	0.03 (0.1%)
Total CI	31,792 *(2.6% of total establishments of MSIC)*	859.9 *(5.7% of total employments in Malaysia)*	29.4 *(1.94% of total national GDP)*
Total	*1,208,875 (as per MSIC)*	*15,073*	*1510.7*

Source: Compiled by authors from DoSM's presentation in Creative Economy 2021 Forum

under four components (i.e. Heritage, Arts, Media, Functional Creations). MCM's classification, however, is in line with the Creative Industry Sector study by Mega Science 3.0 (Akademi Sains Malaysia, n.d.) which has defined creative industries based on four (4) focus areas:

(i) *Heritage*: Galleries, Libraries, Archives and Museums (GLAM)
(ii) *Arts*: Visual Arts, Performing Arts, Traditional Arts and Crafts
(iii) *Media*: Publishing, Audio-visual and New Media
(iv) *Functional Creations*: Built Environment, Fashion and Jewellery, Graphic and Advertising

It is clear that there is no standard classification and taxonomy to define Malaysia's creative and cultural industries. Ad-hoc definitions and classifications are created based on the purpose of a study across different government departments.

4.4 Challenges of the Malaysian Creative City Agenda

The following sections will reveal and discuss some of the blights and challenges in Malaysia's creative city-making agenda.

4.4.1 *Stakeholders' Non-cohesive and Fragmented Efforts*

One of the main challenges in the development of a creative city is the loose and fragmented efforts/initiatives by individuals or by different stakeholders. Instead of working together, many of the creative initiatives are planned and executed individually and separately. The lack of interaction and sharing among individuals/stakeholders will impede the generation of targeted impacts. Understanding stakeholders' perspective is crucial to identify and look into different stakeholder's level of power, influence and interest; the degree of involvement or engagement in certain project; and the legitimacy as well as the relationship among stakeholders (Freeman, 1984; Schmeer, 2000; Friedman & Miles, 2006).

Without collective efforts and effective partnership from each individual/group of stakeholders during all milestones (i.e. planning, development, management, promotion), a creative city may not be sustainably developed. Without a shared vision with the appreciation of shared values in developing a creative city, a collective outcome cannot be achieved thus hampering support and development towards the whole creative ecosystem. Besides fragmented efforts among different stakeholders within one particular sub-ecosystem or within the whole creative ecosystem, there are also fragmented initiatives/efforts with different agendas to develop creative cities in different states in Malaysia. Although the federal government has efforts to promote and facilitate the development of creative industries in the country by formulating the National Creative Industries Policy as well as offer government grants, funding or investments for creative industries players (e.g. investment by MyCreative Ventures Sdn. Bhd[3]; National Creative Industry Grant by Ministry of Information, Communications and Culture, MICC; Content Industry Development Fund[4] by Malaysian Communications and Multimedia Commission, MCMC), such funding information are not effectively distributed among the targeted stakeholders who are in diverse creative fields (MICC, 2010;

[3] MyCreative Ventures Sdn. Bhd was launched in 2012 to manage Malaysia's government investment in local creative companies or establishments through debt or equity investment under the allocation of RM200 million.

[4] Content Industry Development Fund (CIDF-MCMC) program was launched since 2011. It focuses on the development of content for TV, mobile and internet. It aims to develop local creative content industry in line with the national policy objectives under the Communications and Multimedia Act 1998, which targets Malaysia as a global content development hub.

MCMC, n.d.). Some are even unaware of their eligibility in meeting the requirements for the grant's applications. As a result, the federal government's supports have not reached the target groups in different creative fields located in different states/cities within the country. Efforts tend to be individualistic to a particular state where the state government will announce their initiatives in driving the local creative economy. For instance, Kuala Lumpur, Penang, Johor Bahru and Ipoh each attempt to plan and mobilise their own creative and cultural initiatives without involving other Malaysian states.

4.4.2 *Absence of Baseline Data and National Aggregated Statistics*

The initiative to develop the creative economy in Malaysia started since the year 2008–2009 when the first baseline study of Malaysia's creative industries (i.e. Creative Industry Baseline Study 2008 for Multimedia Development Corporation (MDeC) was conducted and the National Creative Industry Policy (NCIP) was formulated. Due to the wide classification for creative industries, the first available baseline findings were not sufficient to reflect the insights of each specific creative field or sector as it merely captured the data and findings for creative content industries, focusing on advertisement, animation, digital content and application, film and TV, games, music and radio (Hasnul, 2011; Hisham, 2016).

The limited baseline data or incomplete national aggregated statistics for Malaysia's creative industries is a serious concern. There is dire need to look into the potential of different creative industries as way to support the actual planning, implementation as well as management of creative industries. Some government-linked institutions have started to map out the local creative industries and drafted development masterplan for creative and cultural sectors. However, those efforts are found mainly in major cities (i.e. Kuala Lumpur, Penang and Johor). For instance, Think City has collaborated with the Kuala Lumpur City Hall, the Ministry of Tourism, Arts, and Culture (MOTAC) and the National Heritage Department to propose the 2019 Kuala Lumpur Creative and Cultural District (KLCCD) Strategic Master Plan. The British Council has produced a report to map creative hubs in Malaysia (British Council, 2017) and Penang Institute regularly reports on matters pertaining to Penang's creative industries.

A significant challenge in collecting accurate data/statistics of creative industries lies in the nature of business operation by creative entrepreneurs in which many of them tend to operate freelance or informally in a small scale. Some of them may not be officially registered. This corresponds with Deuze's (2007) comments about creative industries as an 'hourglass' structure, with few major players driving each creative sector alongside many individual enterprises, small companies and networks of creative talents who are relatively new and working individually. According to Deuze (2007), creative workforce is not highly concentrated but mostly work on 'portfolio careers' where many of them work on multiple jobs across different creative sectors. This scenario has not only constrained a comprehensive coverage in mapping the creative industries in Malaysia, but also challenged the creative workforce in gathering political power and lobbying clout like those of big corporations, established trade unions and traditional large-scale arts organisations.

4.4.3 *Conceptual Confusion and Misnomer*

In Malaysian cities today, noticeably, there are initiatives by local authorities and state agencies to brand their cities as cultural and creative cities. A pioneering endeavour is the capital city of Kuala Lumpur as illustrated in a report entitled 'Kuala Lumpur as a Cultural & Creative City' published in 2018. A study commissioned by Think City, the British Council and other key stakeholders like CENDANA, the report analysed Kuala Lumpur's cultural and creative economy and proposed action plans to develop Kuala Lumpur to be an internationally recognised creative city to stimulate economic growth and development. However, follow-up on progress is slow, especially after the change in Malaysian government in 2020. In due course, other major Malaysian cities like George Town, Ipoh, Johor Bahru and even cities in East Malaysia (i.e. Kuching) are fast hoping on the bandwagon to leverage and showcase their cities' creative and cultural endowments too. However, one may argue that many of these cities have in fact acknowledged and harnessed their respective cultural assets all this while, thus, the creative city label is but a branding agenda and perhaps a conceptual gimmick. This is plausibly true when the agenda is purely neo-liberal urbanism with creative city as a tool and enabler for capitalistic visions and wealth accumulation. False, if, the noble tenets enshrined in UNESCO's Creative Cities Network are fully understood and implemented to include local communities as well as to acknowledge and tap into their traditional skills, local expertise and native knowledge. Past studies have shown that some Malaysian cities tend to lean towards the former when elements of culture, arts and heritage have been indiscriminately commodified as viable tourism goods due to gentrification as in the case of George Town (Foo & Krishnapillai, 2018).

To compound the situation, the lack of knowledge and no common understanding amongst Malaysian urban managers and policy-makers on the definition and functions of a creative city pose another challenge. In this regard, Malaysian policy-makers and urban managers need to fully understand the definition as well as the promise and perils of a creative city first before fully adopting and adapting it to their respective cities.

4.4.4 What Culture? Whose Culture? Whose Creativity?

In exploring the potential of creative city development for Malaysian cities, pivotal enquiries such as '*What culture? Whose culture? Whose creativity?*' inevitably emerged and have to be reckoned with (Mandal, 2008; Jenkins, 2019). For a land rich and endowed with diverse cultures, such questions become more pertinent and pressing because concerns and squabbles often occur with regard to cultural sovereignty. Against Malaysia's backdrop of varied ethnicities and diverse cultures, it raises the enquiry pertaining to which culture is included, excluded or even prioritised as the Malaysian society confront and come to terms with globalisation and its discontent. Unmistakably, a dual layered contestation is at play within the Malaysian scene. The first layer deals with contestations within and between the various ethnicities and cultures thus impacting cultural creativity in all its forms. The second layer, originating from external forces, contends with issues and challenges related to the clash between Malaysian cultural identity and the hegemonic prowess of globalisation. Such an occurrence corresponds to the broader 'clash of civilisations' mantra by scholars since more than two decades back (Huntington, 1997). The turn of the millennium, however, saw the recognition and importance bestowed upon 'cultural diversity' as enshrined in the Universal Declaration of Cultural Diversity 2001 and later in the Convention of the Protection and Promotion of Cultural Diversity in 2005. Such global visions championed the importance of cultural diversity and the critical need to accept and respect diverse cultures across communities and societies. In specific terms, Article 1 of the Declaration perceived cultural diversity as the common language of humanity and that culture can be apparent in a variety of form and manner across time and place. Diversity, thus, is exemplified by the 'uniqueness and plurality of the identities' of the people, communities and societies that compose humankind (UNESCO, 2002: 4).

In a multicultural country like Malaysia, the debates among different ethnic groups on creative initiatives, especially those related to art and cultural subjects, are mostly concerning local cultural roots and identities. The challenge is to reach a consensus among different ethnicities/communities/cultures due to different levels of tolerance and acceptance towards new ideas or people's creativity beyond existing traditions. It poses the contentious question of whose culture should be prioritised as the key to resolve the controversy.

4.5 Efficacy of Creative City Strategy for Urban Sustainability

The following sections will question and deliberate the efficacy of the creative city strategy for urban sustainability in Malaysian cities.

4.5.1 Supply-Driven by State: How Sustainable?

The last decade saw emergence and accelerated momentum of creative city initiatives in Malaysian cities. However, these efforts are unfolding unequally across Malaysia's diverse society. As contested above, the elements of creativity and cultural creativity for that matter have long existed in the Malaysian society or community albeit informally or even in a latent manner – both from a functional and anthropological sense. However, the current situation reveals that advancement in creative city development in Malaysia tend to be selective, economics-driven and heavily dependent on state coffers hence depicting a supply-driven scenario. For instance, the capital city of Kuala Lumpur is a good case in point where key government agencies (i.e. CENDANA, Ministry of Tourism, Arts and Culture) and state-owned urban regeneration establishment (i.e. Think City) perceive the capital city as the 'ultimate creative city to be' with a myriad of international events and funding siphoned there. However, the risk of urban centricity and biasness of concentrating all creative and cultural activities in Kuala Lumpur will exacerbate socio-spatial inequality within and between Malaysian cities and states.

Another example of a state-driven initiative is Iskandar Malaysia in the southern state of Johor where creative industries have been earmarked as one of the nine economic pillars taken by the previous government to transform Malaysia into a high-income, value-added, diversified and innovative economy. Arguably, the state views the creative industry as potentially being able to generate profound multiplier effect to the local economy by not just creating jobs and generating revenue generation for film and television production, but stimulate businesses and create new jobs in other related sectors like retail, hospitality, construction and such. In Iskandar, currently, United Kingdom's Pinewood Shepperton (one of the largest and oldest integrated TV and filming studios and popular for its James Bond movies) is collaborating with Iskandar Malaysia Studios Sdn. Bhd. to jointly develop Pinewood Iskandar Malaysia Studios as operation for an integrated media production studio facility in Nusajaya, Iskandar Malaysia.[5] Launched in December 2013, Pinewood-Iskandar Studios in Johor signalled Malaysia's aspiration to be a regional or even global production location. It is a joint-investment between the Pinewood-Shepperton company (UK) and Khazanah Nasional Berhad (Malaysian government's investment arm) and the construction cost of Pinewood-Iskandar amounted over RM550 million (USD170 million). However, to date, the use of space at Pinewood-Shepperton has been slow with only one feature film, 'Lost in the Pacific' (2015), four television productions by Astro and one theatre production entitled 'P. Ramlee The Musical' (2015) performed in the studio spaces. Hence, Pinewood-Iskandar's aspiration of 'becoming the centre for international film production in Asia' has yet to be materialised. The situation is compounded with regional competition when Infinite Studios Company opened in Batam by Indonesian entrepreneur, Mike Wiluan. Infinite Studios managed to entice and attracted contracts from Singaporean firms as well as from local and international companies (Barker & Lee, 2017: 33).

A similar situation has been observed in Penang too where creative city endeavours tend to be supply-driven with strong state interventions, particularly in funding matters. The Penang state government began to actively invest, support and facilitate the local arts, culture, creative and

[5] Creative Industry in Iskandar Malaysia, published on 22 August 2013. Source: https://propertyiskandar.blogspot.com/2013/08/creative-industry.html (Accessed on 13 January 2020).

heritage initiatives after UNESCO inscription in 2008. The initial aim was to place-make George Town heritage city as a cultural hub by injecting vibrant arts and cultural vibes. For instance, since 2010, the state government has continuously allocated funds for the George Town Festival (GTF), which started as an annual month-long festival and later become a 16-day festival since 2019. GTF aims to celebrate local and international works of art besides promoting appreciation of arts, culture and heritage in Penang. Additionally, Penang state government also invested in establishing the Penang Creative Animation Trigger (CAT) as a global hub for multimedia information, digital content services and creative entrepreneurship (Invest-in-Penang Berhad, 2014). In driving the local art sector as a core sector of creative industries, the state government has announced a budget of RM30 million for the development of Penang Arts District. Such heavy reliance on state coffers questions the sustainability of creative city fanfare in Penang state.

4.5.2 Urban Sustainability

In allaying discontentment related to rising urban inequality and exclusion in creative city discourse (Pratt, 2011), policy-makers and urban managers need to be more critical of the concept by weighing in and addressing the downsides in order to ensure urban sustainability and inclusion. The issue confronting Malaysian creative cities resides in the often naivete roles and myopic visions of political leaders and urban managers who tend to only view and glamourise the economic returns of the concept, thus, justifying the pursuit of a neoliberal urban development pathway. A case in point is capital city of Penang, George Town, which was bestowed as UNESCO World Heritage Site on 7 July 2008. While George Town never received a second UNESCO accolade as a Creative City (of either Gastronomy or Crafts & Folk Art), the rich and diverse cultural endowments embedded within the city's local urban milieu are well-known, unquestionable and could accord the city such a status (Khoo et al., 2014). Very quickly a twist of events unfolded with change in political leadership in 2018 when newly appointed state executive councillor His Excellency Yeoh Soon Hin purported rebranded and changed his portfolio to carry the acronym 'PETACH' which is the abbreviation for Tourism Development, Arts, Culture and Heritage. Extracted from PETACH's website, this move is

'…a refreshed chapter to position Penang competitively in the global tourism map through a state branding under the combination of tourism, arts, culture and heritage in the interest of forging stronger ties and enhance both public and private partnership. At the same time, PETACH is also aligned with the Penang 2030 vision through initiatives under the four portfolios in striving for a smarter and more sustainable state'.[6] It is clear from this statement that 'arts, culture and heritage' have all been lumped together and subsumed under the portfolio of tourism without first understanding clearly the diversity of tourism genres (i.e. heritage tourism, cultural tourism, royal tourism, etc.), their upsides and downsides, and ultimately how these fit into the bigger picture of creative city development especially for a UNESCO historic city like George Town. Recently, in 2020, PETACH has been restructured where 'Arts, Culture and Heritage' are now subsumed under 'Creative Economy' thus the formation of a new acronym 'PETACE' (Penang state exco for Tourism and Creative Economy) (Buletin Mutiara, 2020). The state's understanding of creative economy is rather limited and exclusive to merely 'arts, culture and heritage'. In actual fact, creative economy transcends these fields. This will somewhat create exclusion and challenge urban sustainability. While such challenges to urban sustainability might not be unique to just George Town, other emerging creative cities like Ipoh, Malacca City and Kuching should take heed of the quandaries if aspects of an inclusive and sustainable creative city development were to be championed.

After providing an overview of Malaysia in this chapter, the next four chapters will explore and discuss the viability of creative city as an urban development strategy for Kuala Lumpur, George Town, Ipoh and Johor Bahru.

References

Akademi Sains Malaysia. (n.d.). *Mega Science 3.0: Creative Industry Sector Final Report*. Akademi Sains Malaysia.

Aziz, D. (1987). Perancangan dan pentadbiran kebudayaan di Malaysia. In *Pertemuan Dunia Melayu '82* (pp. 3–40). Dewan Bahasa dan Pustaka.

[6]PETACH. Penang State Exco Office for Tourism, Arts, Culture & Heritage. Source: https://propertyiskandar.blogspot.com/2013/08/creative-industry.html (Accessed on 13 January 2020).

Barker, T., & Lee, Y. B. (2017). Making Creative Industries Policy: The Malaysian Case. *Kajian Malaysia, 35*(2), 21–37.

British Council. (2017, April). *Mapping Creative Hubs in Malaysia*. British Council. Retrieved from: https://www.britishcouncil.my/sites/default/files/malaysia_creative_hubs_report_final_0.pdf. Accessed 18 Nov 2019.

Buletin Mutiara. (2020, October 1). PETACE to Play Pivotal Role in Local Economy, Says Exco Man. *Buletin Mutiara*. Retrieved from: https://www.buletinmutiara.com/petace-to-play-pivotal-role-in-local-economy-says-exco-man/. Accessed 10 Oct 2020.

Deuze, M. (2007). *Media Work*. Polity Press.

DoSM (Department of Statistics, Malaysia). (2019, July 15). *Current Population Estimates, Malaysia, 2018–2019*. Department of Statistics Malaysia.

Economic Planning Unit. (2015, May 21). *Eleventh Malaysia Plan 2016–2020: Anchoring Growth on People*. Economic Planning Unit.

Florida, R. (2002). *The Rise of the Creative Class: And How It's Transforming Work, Leisure, Community and Everyday Life*. Basic Books.

Florida, R. (2004). *Cities and the Creative Class*. Routledge.

Florida, R. (2005). *The Flight of the Creative Class*. Harper Business.

Florida, R. (2007). *The Flight of the Creative Class: The New Global Competition for Talent*. Harper Collins Publishers.

Florida, R. (2008). *Who's Your City? How the Creative Economy Is Making Where to Live the Most Important Decision of Your Life*. Basic Books.

Foo, R., & Krishnapillai, G. (2018). Preserving the Intangible Living Heritage in the George Town World Heritage Site, Malaysia. *Journal of Heritage Tourism, 14*(2), 1–13.

Freeman, R. E. (1984). *Strategic Management: A Stakeholder Approach*. Cambridge University Press.

Friedman, A. L., & Miles, S. (2006). *Stakeholders: Theory and Practice*. Oxford University Press.

Friedman, T. L. (2006). *The World Is Flat*. Penguin Group.

Frost & Sullivan, MyPAA (My Performing Arts Agency), & TFCC (Tom Fleming Creative Consultancy). (n.d.). *Kuala Lumpur as a Cultural & Creative City. A Project Report*. Cultural Economy Development Agency (CENDANA). Retrieved from: https://www.cendana.com.my/clients/Cendana_78A7CADC-1C4A-44E8-A815-E2B4C1D11FE0/contentms/img/Documents/Cendana%20Report%20Web%2006_02_18.pdf. Accessed 18 Nov 2019.

Galloway, S., & Dunlop, S. (2007). A Critique of Definitions of the Cultural and Creative Industries in Public Policy. *International Journal of Cultural Policy, 13*(1), 17–31.

Haji Omar, N. (2016). Message – Menteri Kesejahteraan Bandar, Perumahan dan Kerajaan Tempatan. In *Dasar Perbandaran Negara Kedua (DPN 2)*. Jabatan

Perancangan Bandar dan Desa Semenanjung Malaysia, Kementerian Kesejahteraan Bandar, Perumahan dan Kerajaan Tempatan, Julai 2016.

Hasnul, N. S. (2011, October). *A Snapshot of Malaysia's Creative Multimedia Industries: Trends and Opportunities.* Seminar on 'Enhancing the Competitiveness of the Creative Industry', MATRADE, Malaysia.

Hisham, H. T. (2016). *Sumbangan industri kreatif kepada negara semakin meningkat.* Bacalah Malaysia. Retrieved from: http://www.bacalahmalaysia.com/2016/05/25/sumbangan-industri-kreatif-kepada-negara-semakin-meningkat/. Accessed 13 June 2017.

Huntington, S. (1997). *The Clash of Civilizations and the Remaking of World Order.* Penguin Books.

Invest-in-Penang Berhad. (2014, August 11). *Cashing In on Creative Content.* InvestPenang. Retrieved from: https://investpenang.gov.my/cashing-in-on-creative-content/. Acessed 13 Aug 2019.

Ismail, H. (1990). *Antara Dunia Melayu dengan Dunia Kebangsaan.* Penerbit Universiti Kebangsaan Malaysia.

Jenkins, G. (2019). *Contested Space Revisited: George Town, Penang, Before and After UNESCO World Heritage Listing.* Areca Books.

Kharas, H., Zeufack, A., & Majeed, H. (2010). *Cities, People & the Economy. A Study on Positioning Penang.* Khazanah Nasional Bhd.

Khoo, S. L., Badarulzaman, N., Samat, N., & Dawood, S. R. S. (2014). Capitalising on Urban Cultural Resources for Creative City Development: A Conceptual Review and the Way Forward for Malaysia's George Town. *Geografia Online Malaysian Journal of Society and Space, 10*(5), 20–29.

Kong, L. (2012). Ambitions of a Global City: Arts, Culture and Creative Economy in 'Post-Crisis' Singapore. *International Journal of Cultural Policy, 18*(3), 279–294.

KRI (Khazanah Research Institute). (2018). *The State of Households 2018: Different Realities* (3rd ed.). Khazanah Research Institute.

Landry, C. (2012). *The Origins and Futures of the Creative City.* Comedia.

Lee, C., & Lee, C. G. (2017). The Evolution of Development Planning in Malaysia. *Journal of Southeast Asian Economies, 34*(3), 436–461.

Mandal, S. K. (2008). The National Culture Policy and Contestation over Malaysian Identity. In J. M. Nelson, J. Meerman, & A. R. Embong (Eds.), *Globalisation & National Autonomy: The Experience of Malaysia.* Institute of Southeast Asian Studies.

MCMC (Malaysian Communications and Multimedia Commission). (n.d.). *Term of reference for Creative Industry Development Fund (CIDF-SKMM).* Retrieved from: https://www.mcmc.gov.my/skmmgovmy/media/General/pdf/CIDF_ToR.pdf. Accessed 15 Sep 2019.

MICC (Ministry of Information, Communications and Culture). (2010). *National Creative Industry Policy (NCIP)*. Ministry of Information, Communication and Culture.

Nordin, M., & Hussin, H. (2011). *Malaysian Studies*. Oxford Fajar Sdn. Bhd.

Pratt, A. (2011). The Cultural Contradictions of the Creative City. *City, Culture and Society, 2*, 123–130.

Pua, T. (2011). *The Tiger that Lost Its Roar: A Tale of Malaysia's Political Economy*. Democratic Action Party Malaysia.

Rowland, V. K. (2004). *The Politics of Drama: Post-1969 State Policies and Its Impact on Theatre in English in Malaysia from 1970–1999*. Unpublished M.A. thesis. Singapore: National University of Singapore.

Sasaki, M. (2004). *The Role of Culture in Urban Regeneration*. Forum Barcelona 2004. Barcelona.

Sasaki, M. (2010). Urban Regeneration Through Cultural Creativity and Social Inclusion: Rethinking Creative City Theory Through a Japanese Case Study. *Cities, 27*(2010), S3–S9. Elsevier Ltd.

Sasaki, M. (2011). Urban Regeneration through Cultural Diversity and Social Inclusion. *Journal of Urban Culture Research, 2*, 30–49.

Schmeer, K. (2000). Stakeholder Analysis Guidelines. In *Policy Toolkit for Strengthening Health Sector Reform*. Latin America and Caribbean Regional Health Sector Reform Initiative. Retrieved from: https://www.paho.org/hq/dmdocuments/2010/47-Policy_Toolkit_Strengthening_HSR.pdf. Accessed 15 Oct 2019.

Siddiqui, K. (2012). Malaysia's Socio-economic Transformation in Historical Perspective. *International Journal of Business and General Management, 1*(2), 21–50.

The Borneo Post. (2019, November 21). *New National Cultural Policy to Be Launched in March 2020*. The Borneo Post Online. Retrieved from: https://www.theborneopost.com/2019/11/21/new-national-cultural-policy-to-be-launched-in-march-2020/. Accessed 10 Dec 2019.

Thillainathan, R., & Cheong, K. C. (2016). Malaysia's New Economic Policy, Growth and Distribution: Revisiting the Debate. *Malaysian Journal of Economic Studies, 53*(1), 51–68.

UNESCO. (2002). *Universal Declaration on Cultural Diversity: A Vision, a Conceptual Platform, a Pool of Ideas for Implementation, a New Paradigm* (Cultural Diversity Series No. 1). UNESCO.

World Bank. (1999). *Malaysia Structural Policy Review: Path to Recovery*. World Bank.

World Bank. (2017). Turmoil to Transformation: 20 Years After the Asian Financial Crisis. In *Malaysia Economic Monitor*. World Bank.

Zainuddin, D. (2019, August 13). *MySay: A New Economic Policy for a New Malaysia*. The Edge Malaysia. Retrieved from: https://www.theedgemarkets.com/article/mysay-new-economic-policy-new-malaysia. Accessed 10 Oct 2019.

Zawawi, I. (2000, June). Anthropologising 'National Culture' in Malaysia: Representing and Contesting Culture in the Age of Fragmentation. *Suomen Antropologi, 2*(2000), 50–61.

Zawawi, I. (2004). Globalisation and National Identity: Managing Ethnicity and Cultural Pluralism in Malaysia. In Y. Sato (Ed.), *Growth and Governance in Asia* (pp. 115–132). Asia-Pacific Center for Security Studies.

CHAPTER 5

Kuala Lumpur: Gateway to Malaysia

Abstract This chapter seeks to fathom the viability of Kuala Lumpur (KL) as a potential creative city in Malaysia. This chapter will begin by illustrating the history of KL to set the backdrop. The chapter will also analyse the situation of KL and how the creative city strategy is unfolding in the city. The chapter concludes by discussing the state, private sector and civil society nexus in propounding this strategy.

Keywords Creative city • Urban development • Kuala Lumpur • Malaysia

5.1 Backdrop of Kuala Lumpur

Kuala Lumpur (KL) is the capital and biggest city in Malaysia where it assumes the role as the nation's main commercial, cultural and administrative centre as well as major transportation hub. KL was designated as a municipality in 1972. Subsequently in 1974, this entity and neighbouring areas of nearby Selangor state were declared a federal territory. The land area of the federal territory is 94 square miles (243 square km). Geographically, KL is in west-central Peninsular Malaysia within proximity of Malaysia's Main Range to the north, east and southeast. It experiences equatorial climatic condition with high temperatures and humidity throughout the year (Encyclopaedia Britannica, n.d.). In 2019, KL has an estimated population of 1,776,100, representing 5.43% of the national

S. L. Khoo, N. S. F. Chang, *Creative City as an Urban Development Strategy*, https://doi.org/10.1007/978-981-16-1291-6_5

population (DoSM, 2020a). As a federal territory and the capital city of Malaysia, KL has the highest population density in Malaysia (DoSM, 2019). As per 2019 data, a household in KL earns a median monthly income of RM10,549 and mean monthly income of RM13,257 (DoSM, 2020b), which is much higher if compared to other states. The incidence of absolute poverty for KL has improved to 0.2%. (DoSM, 2020b). Kuala Lumpur's economy relies mainly on services sector which contributes 87.8% to its GDP in 2018 (DoSM, 2020a).

KL has its origins as a mining town. Although there were early settlers in the area before middle of the nineteenth century, KL was only formed as a settlement in 1857 when Raja Abdullah and Raja Jumaat brought in a group of 87 Chinese miners. These Chinese immigrants landed at the confluence of the Klang River and Gombak River. The community thrived as a tin-collecting centre and subsequently the area morphed into a commercial post and mining settlement known as KL – which literally means 'muddy confluence' in Malay. Later in 1880, KL outshone Klang as the state capital and further flourished under British resident Sir Frank Swettenham. Its strategic central position accorded it the ideal choice as capital of the Federated Malay States.

KL was not spared from Japanese occupation (1942–45) during World War 2. After the war, KL's population grew dramatically and in 1957 it became the capital of the independent Federation of Malaya and of Malaysia in 1963. The post-war era or Merdeka era marked a defining moment for Malaysia and capital city KL where the nation was developing and modernising itself (TCSB, 2019: 22). At that point, Malaysia as a new nation-state focused on industrialisation to develop the country. Growth ensued in the following years followed by rapid urbanisation. The following years saw huge exodus of rural to urban migration taking place, especially to big cities like KL. By mid-1960s, the population in KL soared to half a million and eventually hit the one million mark in the early 1980s. Increased population eventually started to congest the city. The way Malaysian government offices were scattered across the city caused disruptive administration. Eventually, many federal offices in KL were relocated to Putrajaya, a newly planned city south of KL. This shift took place during the turn of the twenty-first century. Subsequently, KL was eased of administrative duties when Putrajaya assumed the role as seat of administration and KL remained as capital city (Encyclopaedia Britannica n.d.).

Right after World War 2, KL's built environment saw the sprouting of iconic civic structures of Merdeka (Independence) architecture as a way to construct national identity and patriotism. The 1980s and 1990s were milestone eras where KL experienced significant changes, rapid urban development and modernisation with the completion of gargantuan infrastructural projects. However, rapid growth and urban sprawl due to development of 'new centralities' on the city's outskirts were the root cause of the hollowing out process and depopulation of the historic inner city of KL (TCSB, 2019: 22).

The late 1980s and 1990s saw Malaysia heralding globalisation and neo-liberal reforms in a big way. Under then premier Tun Mahathir who set the tone for neo-liberalism (since the 1980s), the overtones of laissez-faire economy were quickly permeating into all realms of development, including urban development. KL quickly set an aspiration to be a global city with increasingly global orientation of the city displayed through material infrastructure and spectacular symbolic spaces (Bunnell et al., 2002). However, research has shown that while changes affecting KL's built environment (due to its integration with the global economy) has successful attracted capital flows into its built environment, but, it has not reacted well to globalisation in terms of overall planning and growth management (Sirat, 1997). To address this, KL seeks to achieve equilibrium between physical, economic, environmental and social sustainability as envisioned in their 'Kuala Lumpur City Plan 2020 Towards a World Class City' (Kuala Lumpur City Hall, 2018).

However, subsequent actions that followed suit included massive and copious physical development in the late 1990s planned for Greater KL and further afield. All these new developments impacted the historic core in one way or another. For instance, the transformation of Bukit Bintang into a new Central Business District, relocation of administrative offices (i.e. Prime Minister's Office and Law Courts) to Putrajaya, construction of new national stadium at Bukit Jalil as well as the emergence of new suburban commercial and retail hubs have all drawn population away from the historic core of KL (see Fig. 5.1). The population size of KL City Centre (bounded by the middle ring road) dwindled from 130,000 in 1990 to below 100,000 by 2010 (TCSB, 2019: 22). A recent study by Think City in their report '*Kuala Lumpur Creative and Cultural District Strategic Master Plan*' has fleshed out all these issues and discussed how the historic core of KL is facing a dearth of population, low economic diversity and lacks a core economic function (TCSB, 2019: 22). For the purpose of this book, the boundary for KL under study would be the historic core area akin to Think City's 2019 study.

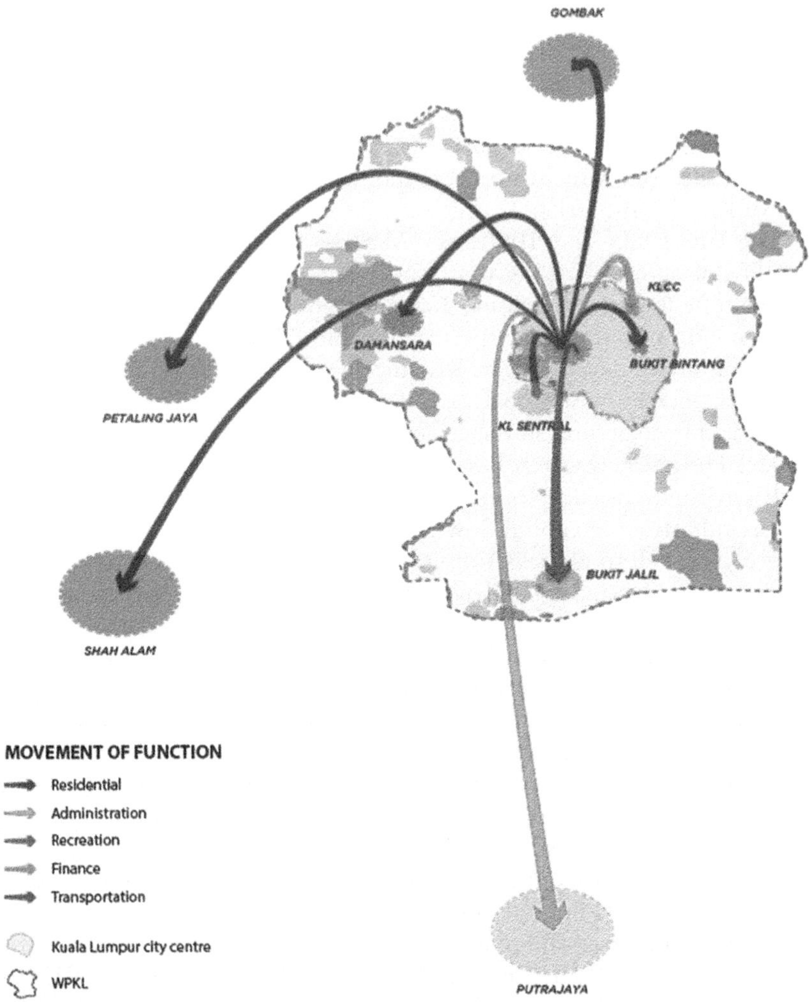

Fig. 5.1 Location of KL historic core and the hollowing out process. (Source: TCSB, 2019: 28)

5.2 Situation Analysis of Creative City Strategy for KL

The next sections will analyse the situation of KL and how the creative city strategy is occurring in the city.

5.2.1 *Advantage as Capital and Primary City in Malaysia*

The historic core of KL still retains the aura, advantage and charm as capital city in Malaysia. Geographically, its midpoint location in the centre of Peninsular Malaysia is strategic and convenient for other Malaysian secondary cities like Ipoh, George Town, Malacca city to converge and transit. Apart from the city centre's rich historical, cultural and architectural significance, KL city centre is inevitably the main gateway and entry point to Malaysia. This is made possible by a modern transportation system like the ERL and shuttle services from KLIA1 and KLIA2 to *KL Sentral* which is a transit-oriented development located in the heart of KL. Additionally, KL is also the only Malaysian city that has an integrated public transport system operated by My Rapid that oversees the KL Monorail Line and LRT lines for Kelana Jaya, Sri Petaling and Ampang. Besides that, KL also boasts a wealth of creative and cultural assets within its urban milieu. Key national icons and civic landmarks like the National Museum, National Mosque, KTM Office, *Medan Pasar*, Central Market, Brickfields and colonial buildings adjoining to *Dataran Merdeka* are valuable cultural endowments that can be harnessed to build KL's creative and cultural base. As espoused in creative city literature (Landry, 2008), a viable creative city is attributable to the availability of both the hardware (i.e. buildings, roads, waterways) and software (i.e. people, history, talent). In this case, the historic core of KL has its fair share of hardware in the form of old colonial buildings such as the shophouses around *Jalan Tun H.S. Lee*, *Medan Pasar* and the inner city, that can be restored, refurbished and adaptively reused. However, during a recent site visit to the inner city in 2019, many of these shophouses are old, derelict, dirty and somewhat pose as an urban eyesore to the public.

While urban regeneration think-tanks (i.e. Think City) have endeavoured to find ways to regenerate inner city KL and pockets of success are evident like the *Medan Pasar* area, the next question would be the viability of urban sustainability which is contingent upon financial and funding continuity given that projects and hand-outs from Think City are funded from Khazanah Nasional Berhad, which is the investment arm of the Malaysian government. Over-dependency on state's coffers to run such initiatives is not sustainable in the long term. However, it is heartening to discover through fieldwork interviews that private initiatives as well as jointly funded Think City and private endeavours are currently happening to refurbish dilapidated shophouses and properties in the inner city for creative use. One good example of a jointly funded project by Think City and private owners is the Zhongshan Building where an old apartment block was refurbished and literally brought back to life. While inner city KL still possesses substantive cultural assets to warrant it a viable creative and cultural district of sorts, systematic and concerted efforts by all key stakeholders are vital and they need to be on the same platform to achieve common goals. Notwithstanding, the birth of a '*Kuala Lumpur Creative and Cultural District Strategic Master Plan*' by Kuala Lumpur City Hall and Think City in collaboration with the Ministry of Tourism, Arts and Culture (MOTAC) and Department of National Heritage is much welcomed, but the implementation and fruition of this plan is what matters most. This can be a challenging feat given the recent change in Malaysian government in 2020. Change in ministerial portfolios would affect the buy-in and implementation of this development plan subject to the new ministry's understanding and acceptance of the creative city model as an urban development strategy.

5.2.2 KL – Magnet for Creative Talent?

As an aspiring world-class city, inner city KL and its adjoining areas like *Bukit Bintang*, *Jalan Tun Perak* and *Jalan Yap Kwan Seng* are considered as pseudo-Wall Street financial districts, high-end commercial hubs and cultural precincts for KL. In fact, in previous studies undertaken during the 1990s and the turn of the century, KL was already aspiring to brand itself as a world city (Sirat, 2000). These amenities and creative/cultural avenues with a global outlook are in fact places, spaces and hangouts that appeal to the 'creative class' as purported by Florida's econo-centric model though a big portion of the retail establishments are high-end global brands housed in 5-star malls (i.e. *Starhill*, *Pavillion*) that offer global luxury brands like *Coach*, *LV*, *Burberry* and such – thus raising the question of production and consumerism that caters to whom and what cohort of consumers. Obviously, the working-class locals would not be patronising these establishments.

Nonetheless, the key enquiry would be to find out whether the availability of such state-of-the art physical amenities can transform KL into a magnet for creative talent. The key informant interviewees acknowledged KL's capital city status, international outlook and easy accessibility to a lot of funding, institutions, venues and potential audience. Thus, KL is in fact starting with an advantage compared to other Malaysian cities. However, when prompted to comment on growth of creative talent in the city, their responses were less promising as stated here:

> We should be playing a bigger role but we're not. Infrastructure and qualified people are missing. (Creative hub owner, interview, 24 July 2019)

> Lack of growth. If you're entering for the first time…the fresh feeling…I'm someone new. After that, after your 5th year, is there a future for me here? Space on multiple levels…mental level, freedom, rights level, monetary level…it feels as if it is staying the same size even as you are trying to grow individually with your art. A lot of people are getting more recognition when they leave. (Programme Officer, International organisation for cultural relations & education, focus group discussion (FGD), 22 July 2019)

The situation is compounded when the elements of 'creativity and the creative minds' were never really tapped into by Malaysia as a nation. As a result, there is a dearth of talent in Malaysia's creative fields. Worse still, those who have succeeded are reluctant to share and nurture new blood.

> As a nation we didn't tap into our creative minds for a long time. There are only a few key players. You see the same key players everywhere, so it feels as if it has been maxed out, but it is not. There are other people who can run content, but they are still finding their way. Even for those over-exposed ones, they do the same thing, they go for the similar circles and they have gone through the hardship and they don't want to share. The problem is we have that silly mindset that … "I went through this, so you should go through this." It shouldn't be. (Urban dweller-cum-artist, interview, 24 July 2019)

5.2.3 A Culturally Well-Endowed Inner City

Just like most inner cities, KL historic core is not short of cultural endowments seen through its unique architecture, history, creative and cultural enterprises. The '*Kuala Lumpur Creative and Cultural District Strategic*

Master Plan' mapped out the city's creative and cultural industries as featured in Figs. 5.2 and 5.3 below. There are great potentials in these assets if KL intends to be a creative city. In terms of built environment, KL boasts a host of traditional shophouses with *Straits Chinese Eclectic* style, *Art Deco* transformation plus traditional government and commercial buildings with *Moorish*, *Neo-Classical* and *Art Deco* styles.

These historic landmarks, nonetheless, are not spared from the pressures of modern development in the city. The impacts of global restructuring to the socio-spatial aspects of KL inner city are distinctively felt now (i.e. 2019) though early observations by scholars (Bunnell et al., 2002; Sirat, 1997, 2000) have evidenced such emerging socio-spatial changes since two decades ago. To date, while KL historic core has mixed land use comprising of mostly commercial entities (business activities, retails, offices, wholesale/warehouses, gastronomy, etc.) but the city's traditional building form and function of shophouses are rapidly vanishing (TCSB, 2019: 39). Interviews with key stakeholders disclosed that many of these shophouses in the inner city fell into disrepair, were abandoned, derelict

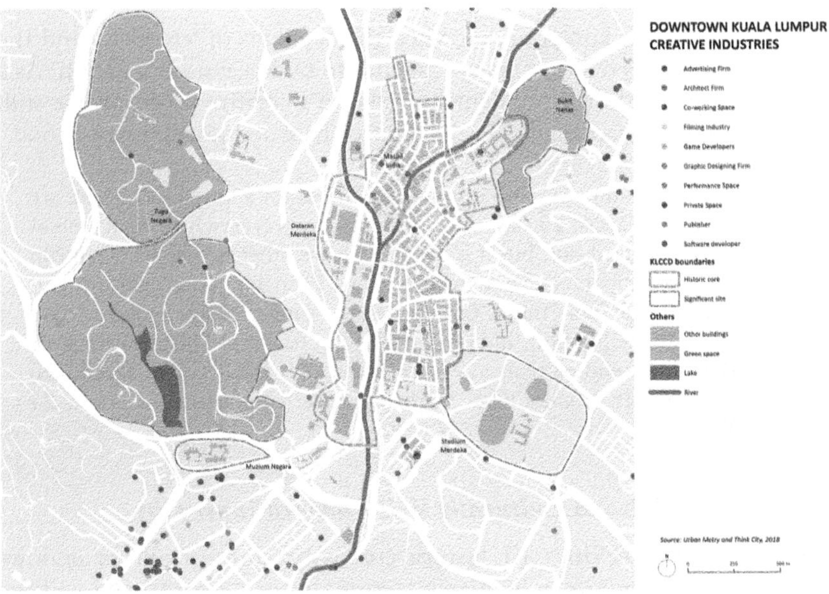

Fig. 5.2 Creative industries in downtown KL. (Source: TCSB, 2019: 65)

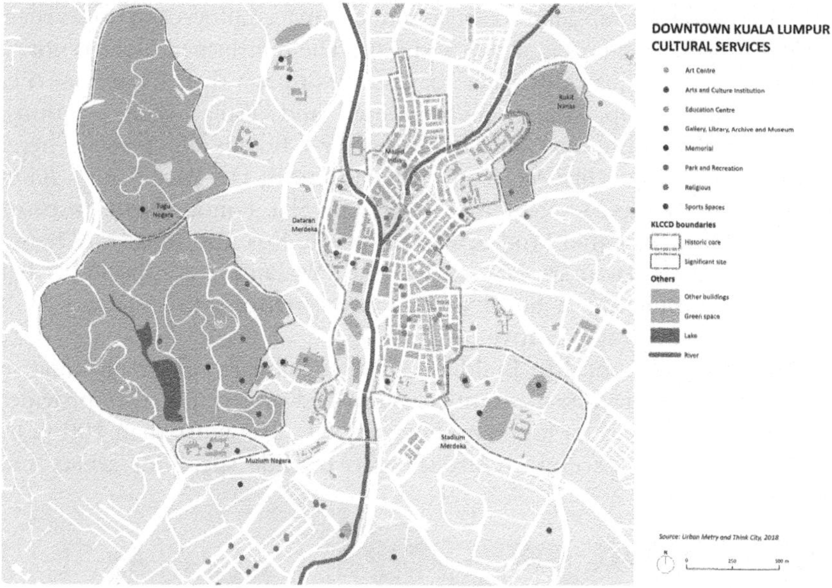

Fig. 5.3 Cultural services in downtown KL. (Source: TCSB, 2019: 64)

and became haunts for the homeless. However, in the last few years, revitalisation activities were gradually mushrooming. Interestingly, these are local and private initiatives started by the second generation of existing landlords of these shophouses. Old buildings are revived and adaptively reused. One sterling example is REXKL which was a cinema that was burnt down twice. A group of young architects leased the space and within their capacity and expertise, they revived the space and activated it as a creative hub. More importantly, REXKL is designed with a novel purpose to allow people to create. As highlighted by the design architect Chang, "When we say that we are a space for people to create, we want to emphasise on the full spectrum of what is involved in the creative process. This means finding the means and resources in order to achieve the execution of one's creation. And that's the REXKL community. Bringing people together to create." (Wong, 2019).

As mentioned earlier, another key exemplar is the Zhongshan Building which was an old and derelict block of flats owned by a local KL family. With funds from Think City, the owners managed to refurbish the building and later sub-let it to artists and creative employees to use as their SOHO spaces (see Figs. 5.4, 5.5 and 5.6). The refurbished block is now a creative hub that provides space for creatives to run their businesses (i.e. gallery, book shop, studio, archive, etc.) and display their creations. Interestingly, non-global and non-designer's brands are present here. Homegrown talents like Ana Tomy, Atelier Fitton, The Back Room, *Balai Buku Raya*, Bogus Merchandise, Studio Ejin Sha, Fono, *Huruf*, *Institute Pondok Perancis*, Kanid Studio, *Kanta Studio*, Kok Far, Malaysia Design Archive, Muhendaran Sri, Naiise Malaysia, *Piu Piu Piu* are spotlighted at Zhongshan Building, a novel way to acknowledge and showcase KL's local and homegrown creative talents.

Apart from Zhongshan Building, another fully private sector initiative is the *Godown* project, a recently refurbished arts centre that provides space for artists and art lovers in inner city KL. The more than 100-year-old building was recently refurbished to provide space for artists and

Fig. 5.4 The newly refurbished 12-unit Zhongshan Building. (© Suet Leng Khoo (July, 2019))

Fig. 5.5 Interior of Zhongshan occupied by a café. (© Suet Leng Khoo (July, 2019))

curators to work and host workshops/exhibitions (see Figs. 5.7, 5.8 and 5.9). Such initiatives will in one way or another gentrify KL inner city.

While gentrification is generally understood as the coming of gentrifiers to urban areas and concurrently the arrival of new forms of businesses and cultures that do not normally sync with local settings (Foo & Krishnapillai, 2018), in KL inner city, the gentrification process seemed to have a twist. Gentrification is subtly welcomed in certain areas. This is mentioned by one local artist who has resided in KL for the past 10 years:

> In this case, gentrification is very slow, subtle and it actually might help because... I'm scared to say that word. But in this case, it is needed. It is the younger generation here in KL. They know the pros and cons of gentrification and they control it as well. They are more in the know about what's happening. What's right and wrong. They are educated in that way. They know when people are oppressed and not oppressed. They are more well-

Fig. 5.6 Art gallery at Zhongshan building. (© Suet Leng Khoo (July, 2019))

> informed. When they see gentrification happening, they will be up in arms. (Urban dweller-cum-artist, interview, 24 July 2019)

With more 'youthfulness' and vibrancy re-entering KL inner city through such gentrification efforts, the option of harnessing KL's creative energy and culture has its potential if planned and managed properly. To date, national and state initiatives that acknowledge KL's role is visible through the report of '*Kuala Lumpur as a Cultural & Creative City*' (Frost & Sullivan, MyPAA and TFCC, n.d.). The KL City Hall (via its Cultural Department) also has a line-up of creative endeavours like the KUL Month and Art in the City initiatives. More recently, Think City's '*Kuala Lumpur Creative and Cultural District Strategic Master Plan*' are commendable efforts provided they are carried through till the end. This is quite a cause for concern given that Malaysian development plans have this tendency to remain as rosy public blueprints that merely remain as documents without running its practical course. In the masterplan, the

Fig. 5.7 The exterior of the *Godown* – new arts space in KL. (© Suet Leng Khoo (July, 2019))

collective decision and strategy by KL City Hall, Think City (urban regeneration think-tank) and the Department of National Heritage (JWN) to adopt UNESCO's Historic Urban Landscape (HUL) integrated heritage management tool to steer and manage heritage conservation and urban development in KL's vibrant and fast changing environment does show the leaning of KL towards a culture-led urban regeneration and creative city development as espoused by Landry (2008). Having understood the present scenario, it is still crucial to ensure that gentrification of the physical environment should not be at the expense of original and local communities, thus, ensuring a socially inclusive creative city.

5.2.4 *Non-cohesive Creative and Cultural Ecosystem*

During the interviews, one of the key issues highlighted was the non-cohesive creative and cultural ecosystem in KL. Whilst many of the endeavours (i.e. heritage trails by KL City Hall, grassroots arts initiatives) thus far

Fig. 5.8 A scene from *Godown* looking out to KL old town. (© Suet Leng Khoo (July, 2019))

are in fact creative initiatives and are instrumental towards the making of a creative city, ironically, the fragmented and non-cohesive ways that these initiatives were undertaken will undermine the growth of the creative industries and its ecosystem in KL. The quotes below explain the scenario.

> Cohesive is a hard thing for KL…KL does not do well with cohesion. Obviously, there are grassroots to mobilise creativity, but I find them often to be very small scale. In KL, the amorphousness of the city. We're KL, we're central. Sometimes artists becoming very hyper-local and they become very aligned to their neighbourhoods and they become…I'm not of KL…but I'm of *Bangsar* or *Kampung Baru*. To directly answer this question, I don't think it is strong because it isn't cohesive. It's the same issues of silos, inconsistency. (Programme Officer, International organisation for cultural relations & education, FGD, 22 July 2019)

Fig. 5.9 Interior of *Godown* that can be used to host events. (© Suet Leng Khoo (July, 2019))

> I don't know whether it is a consequence or cause. Everybody ends up working in silos. The creatives themselves are not connected. It is all levels that are disjointed. (Head of Arts & Creative Industries, International organisation for cultural relations & education, FGD, 22 July 2019)

The disjointed nature has inadvertently caused KL's ecosystem to suffer from discontinuity in actions taken and thus hamper sustainability. Before an action plan can run its full course to fruition, key informants highlighted that it would halt midway and the same idea will be undertaken again by a separate party right from the start and it is akin to hitting the *"reset"* button. This scenario was lamented by key informants from KL's renowned arts and culture advocacy institution during a focus group discussion:

> It is very disjointed, and we always must click the 'reset'. There's a lot of fire-fighting mentality and reactionary mentality. You're reacting to differ-

> ent changes because our government changes every 5 years. People change within the 5 years and the Ministry of Tourism, Arts and Culture (MOTAC) and their mandates change every 5 years ... What's next after the event? How can we support whoever is inspired or sparked or moved to participate in the city? There's no follow-up and no plan. As someone who organised these events, I'm also guilty. Let's just do the workshop. People get into one space and do and after that there's no follow through. A lot of people would be walking away after that and not having anything concrete. There's this pattern of stopping and petering out. All we're having is a collection of first steps. No second or third step. (Programme Officer, International organisation for cultural relations & education, FGD, 22 July 2019)

> I also feel that is happening...if you're talking about the govt initiatives, very often it is not followed up. The conversation happens and there's no follow up. That's what you said earlier. The reset button. And what that creates? It creates '*blasé*' attitude...not another of those again. They don't believe that things can change. (Head of Arts & Creative Industries, International organisation for cultural relations & education, FGD, 22 July 2019)

5.2.5 *Social Exclusion*

Besides issues within the creative and cultural ecosystem itself, the broader city structure upon which the industry and sub-sectors positions itself warrants attention too, especially when elements of social exclusion are continuously plaguing KL's inner city amid endeavour to transform the city into a creative and cultural district by state authorities. From the interviews and fieldwork visits in July 2019, it is evident that KL historic city is quite hollowed out and devoid of local residents apart from working hours. While KL historic core still has mixed land use comprising mostly commercial activities (i.e. business activities, gastronomy, retails, offices, etc.), but, the city's original urban fabric, traditional building form (i.e. shophouse) and its functionality are quickly vanishing (TCSB, 2019: 39). From primary and secondary data collected, it is noticeable that KL inner city has two (2) huge vulnerable groups, namely, (i) homeless people, and (ii) foreign citizens who are making a marked presence working, living and spending leisure time in the city. Key informants disclosed that a huge stock of shophouses in inner city KL are derelict, abandoned and became haunts for the homeless. Recent findings reported that KL historic core has approximately 300 homeless people in the area at night (TCSB, 2019:

61). Despite the hype of KL as a world class city and now this acid test of its viability as a creative city, the presence of homeless people and their disadvantaged role in the city is real and needs to be reckoned with (see Figs. 5.10 and 5.11). The scenario was disclosed by a local urban dweller-cum-artist who lived in KL for the past 10 years:

> As a city dweller, I feel the old part of the city dies down after 6–7 pm which means locals and foreigners have nothing much to do. The roads feel a bit overwhelming and scary because they are dark and they are inhabited and there are many, many homeless people. If you were to come back here tonight at H.S. Lee Road and the other road, you'll be shocked because the amount of lunchtime footprint on a weekday is similar as the amount of homeless at night just two streets away. I thought there was an event happening. I realised they are homeless people and they came out from every nook and cranny. (Urban dweller-cum-artist, interview, 24 July 2019)

Fig. 5.10 Homeless people around *Medan Pasar* area. (© Suet Leng Khoo (July, 2019))

Fig. 5.11 Homeless people and vagrants in downtown KL. (© Suet Leng Khoo (July, 2019))

Besides the homeless, the migrant community is also facing the brunt of social exclusion. The depopulated KL historic core now has a meagre residential population of just 1000 where about half are foreigners who are mostly males. The main underlying issue is the discordance between jobs, housing type and its quantity. KL is presently 'home' to a sizeable migrant community mostly from Bangladesh, Nepal and Myanmar who are in KL searching for a livelihood or to visit. KL's limited public realms are converted into meeting and hang-out spots for these communities, especially over the weekends (TCSB, 2019: 61). By just strolling KL inner city in the morning, visual observations affirm that the historic core is home to a huge migrant community.

The concern now would be what have been done thus far to include these groups of people into KL's creative city development. Scholars like Sasaki (2011) advocated that creative cities need to be socially inclusive to be sustainable where the 'creative and innovative milieus' in cities should be respected and grassroots capabilities need to be collated to address social problems such as homelessness. Evidence from the ground signals that as KL aspires to be a creative and cultural city, the whole approach adopted thus far is seemingly 'capitalistic' and the socio-spatial inequality is real whereby the interests of vulnerable groups (i.e. homeless people, migrant workers) are rarely in the equation as quoted here:

> Gentrification is happening because of the spaces that are occupied. So much reclamation of space especially in a city as old as KL. There's that trend of revitalisation spaces by using it creatively which I'm not against. I think it is great in KL and George Town, they are doing them in a cool way. But the approach is still very capitalistic, and it is about commodification … the example of the art murals in George Town … We're talking about programs. We had elements like…is it possible to make things Instagrammable? There are these considerations that didn't exist 10–15 years ago that is changing the landscape of what we expect to see in a creative space let alone a creative city. That I think glosses over the creative city. Where are the migrant workers, where are undocumented people in the creative city? (Programme Officer, International organisation for cultural relations & education, FGD, 22 July 2019)

Such evidences of social and urban exclusion have challenged the notion of cultural diversity as spotlighted in the Universal Declaration of Cultural Diversity in 2001, and later the Convention of the Protection and Promotion of Cultural Diversity in 2005. While these global visions acknowledged the importance of cultural diversity as the 'common heritage of humanity' and the need to respect diverse cultures across societies, in KL's case, there is little recognition of the presence of foreign labour, their cultures and their creative talents thus connoting KL's creative city aspiration less socially inclusive. The question of inclusion or exclusion of a foreign culture in the local context, indisputably, will open a Pandora's box pertaining to concerns that cultural anthropologists have long sought to address such as '*Whose culture?*' and how identity and authenticity of communities/places are negotiated to accommodate this new phenomenon (Jenkins, 2019: 3) which then leads to the legitimacy and rights of foreign labour and their urban citizenship. To ensure a sustainable pathway, KL's grand idealism to be a creative and cultural city/district will first need to come to terms and fathom the changing urban demographics and dynamics within the city centre with influx of migrant workers and the presence of vulnerable groups where some are the invisible labour force and are integral to fuel the informal urban economy or assume essential occupations in the lower rung of the occupational hierarchy. Without these 'invisible' army of foreign labour (i.e. drivers, janitors, canopy installers, cleaners, etc.), any big creative event (i.e. shows, concerts, etc.) would not be possible let alone the development of a creative city.

5.2.6 *Failure of Education System to Nurture Creativity*

Besides the overall city structure, the complementary component of human agency is also important. As espoused by Landry (2008), a creative city entails both hard and soft infrastructure. In KL's case, while Greater KL boasts physical structures and urban amenities comparable to a world class metropolis, the city is experiencing a dearth of human capital to fuel creative and cultural industries. CENDANA's 2018 report highlighted that the shortage of human capital was inextricably linked to Malaysia's education system and societal mindset that failed to nurture creativity, innovation, arts and culture. To exacerbate the situation, KL is confronted with issues of brain drain due to insufficient audiences to support the industry and issues related to freedom of expression and censorship. The interviews highlighted the magnitude of the education issue as cited below:

> My own interpretation is it starts at the education system where arts, creative thinking and creativity are not a priority. The emphasis has been on technology, science and on the digital economy. The arts take a backseat. It is not encouraged to some degree. There are probably political reasons for that. (Programme Officer, International organisation for cultural relations & education, FGD, 22 July 2019)

> We all focused on Dr. Mahathir's vision to be technologically advanced and developed, and we left all the arts and creativity behind. I always look at Indonesia, Thailand and Philippines as pure. In *Asia Bagus*, who wins the contest? Malaysia does not win, the 3 wins. Singapore has the economic support to be creative and strategic and not so many religious rules that constrain. (Urban dweller-cum-artist, interview, 24 July 2019)

> Our education does not nurture and Malaysians don't learn to appreciate. There is little appreciation for arts and culture. (Creative hub owner, interview, 24 July 2019)

5.3 The Nexus Between State, Market and Civil Society

In order to understand the development and future of KL's creative and cultural industries which underpins the creative city concept, the roles, relationships and dynamics between the state, market and civil society are important. In KL, insofar, there are existing efforts to nurture the creative and cultural industries but noticeably these are top-down initiatives as seen through the works and interventions by Think City, KL City Hall and CENDANA which are all government-linked and public-funded institutions to regenerate cities and develop the arts and cultural sectors respectively. This approach, however, does signal a supply-driven scenario where KL's creative and cultural industry thus far is rather reliant on public coffers although there are pockets of private sector initiatives like *Godown* and REXKL emerging. The way forward would be to motivate more public-private-civil society strategic partnerships with a strong leaning towards community-based approaches. Ultimately, a city, whether creative or not, must be people centric. Hence, community participation, inclusion and empowerment of the entire KL urban citizenry is of utmost importance.

In KL, while there are existing grassroots movements in its creative and cultural industries, the efforts are fragmented and disconcerted which resulted in a disjointed ecosystem. Redundancies and overlaps in creative efforts by artists and cultural workers where they constantly redo and restart a similar type of project (as previous ones) have somewhat impeded the growth of the sector in KL. By doing the same thing repeatedly does not move the industry forward but stagnates it instead. Additionally, the 'soloist, independent contractor and freelance' work nature of creative and cultural employees are in part the cause that breeds non-cohesion and disjointedness in the industry despite artists arguing that their works are akin to masterpieces and going solo is the way to build reputation and etch their hallmark of distinctiveness.

To compound the matter, just like many other econo-centric creative cities, stakeholders in KL's creative and cultural industries are either ignorant or oblivious towards the actual definition, classification, upsides and downsides of the creative city concept. The situation is further aggravated when arts and cultural employees lamented that arts or cultural products in KL should not be loosely quantified and carelessly commodified to meet the performance yardstick of the Ministry of Tourism, Arts and Culture, which is to boost tourism and to meet economic ends. The acronym MOTAC itself which has the tourism component embedded alongside arts and culture can take a wrong turn and go awry if commodification of culture and creativity lean excessively to be econo-centric rendering it less inclusive and more elitist.

References

Bunnell, T., Barter, P. A., & Sirat, M. (2002). KL Metropolitan Area: A Globalizing City-Region. *Cities, 19*(5), 357–370.

DoSM (Department of Statistics Malaysia). (2019, October). *Malaysia Statistical Handbook 2019.* Department of Statistics, Malaysia.

DoSM (Department of Statistics Malaysia). (2020a, February). *Poket Stats Negeri Wilayah Persekutuan ST4 2020.* Department of Statistics, Malaysia.

DoSM (Department of Statistics Malaysia). (2020b, July). *Household Income & Basic Amenities Survey Report by State And Administrative District – Wilayah Persekutuan 2019.* Department of Statistics, Malaysia.

Encyclopaedia Britannica. (n.d.). *KL.* Encyclopaedia Britannica online. Retrieved from: https://www.britannica.com/place/Kuala-Lumpur#accordion-article-history. Accessed 9 Apr 2020.

Foo, R., & Krishnapillai, G. (2018). Preserving the Intangible Living Heritage in the George Town World Heritage Site, Malaysia. *Journal of Heritage Tourism, 14*(2), 1–13.

Frost & Sullivan, MyPAA (My Performing Arts Agency) and TFCC (Tom Fleming Creative Consultancy). (n.d.). *Kuala Lumpur as a Cultural & Creative City.* A Project Report. Cultural Economy Development Agency (CENDANA), Malaysia. Retrieved from: https://www.cendana.com.my/clients/Cendana_78A7CADC-1C4A-44E8-A815-E2B4C1D11FE0/contentms/img/Documents/Cendana%20Report%20Web%2006_02_18.pdf. Accessed 18 Nov 2019.

Jenkins, G. (2019). *Contested Space Revisited: George Town, Penang, Before and After UNESCO World Heritage Listing.* Areca Books.

Kuala Lumpur City Hall. (2018). *Kuala Lumpur city Plan 2020 Towards a World Class City.* Kuala Lumpur City Hall.

Landry, C. (2008). *The Creative City: A Toolkit for Urban Innovators* (2nd ed.). Comedia Earthscan.

Sasaki, M. (2011). Urban Regeneration Through Cultural Diversity and Social Inclusion. *Journal of Urban Culture Research, 2*, 30–49.

Sirat, M. (1997). Globalization of Economic Activity and the Changing Landscape of Kuala Lumpur Central Planning Area. *Asian Geographer, 16*(1–2), 37–58.

Sirat, M. (2000). Globalising Kuala Lumpur and the Strategic Role of the Producer Services Sector. *Urban Studies, 37*(12), 2217–2240.

TCSB (Think City Sdn. Bhd.). (2019). *Kuala Lumpur Creative and Cultural District Strategic Master Plan*. Prepared by Kuala Lumpur City Hall and Think City in Collaboration with Ministry of Tourism, Arts and Culture Malaysia and National Heritage Department. Think City Sdn. Bhd.

Wong, L. Z. (2019, November 18). *REXKL, an Old Iconic Cinema Reborn into a Unique Arts and Culture Hub*. The Star Online. Retrieved from: https://www.thestar.com.my/lifestyle/living/2019/11/18/remember-the-old-rex-cinema-on-jalan-sultan-kl-it039s-been-reinvented-as-rexkl-a-bustling-arts-culture-and-events-space. Accessed 16 Apr 2019.

CHAPTER 6

George Town – UNESCO World Heritage Site

Abstract Apart from Kuala Lumpur, another Malaysian city that is embarking on the creative city pathway is George Town. This chapter illustrates how the creative city agenda is playing out in George Town World Heritage Site, especially after the 2008 UNESCO inscription. The chapter will close by discussing the state, private sector and civil society dynamics in the process.

Keywords Creative city strategy • Urban regeneration • Urban development • George Town • UNESCO World Heritage Site

6.1 Background of George Town

George Town (GT) was founded by Captain Francis Light from the British East-Indian Company (EIC) in 1786. It then emerged as a free port town during the British colonial era (Encyclopaedia Britannica, n.d.). It also flourished as a busy capital and commercial centre of the Straits Settlements. Despite Singapore replacing Penang as the capital of the Straits Settlements in 1832, George Town retained its position as a main British entrepot. In 1867, the Straits Settlements became British Crown territories. Under direct British rule, George Town gradually expanded with improved public transportation. George Town's rapid development has made Penang island a successful and strategic trading hub for northern Peninsular Malaysia and the Indonesia-Malaysia-Thailand Growth Triangle in the

S. L. Khoo, N. S. F. Chang, *Creative City as an Urban Development Strategy*, https://doi.org/10.1007/978-981-16-1291-6_6

early days. Traders and travellers from Asia and Europe stopped by Penang island and congregated at George Town for business (Turnbull, 2002). By the end of the nineteenth century, George Town evolved into a leading financial centre in Malaya where established international banks such as Standard Chartered and HSBC (Hongkong and Shanghai Banking Corporation) began to cluster at Beach Street. However, Penang's development was disrupted during the Japanese occupation from 1941 to 1945 (Langdon, 2014). In 1946, the British re-occupied the island and dissolved the Straits Settlements. This was followed by merging the Crown Colony of Penang into Malayan Union which was later replaced with Federation of Malaya in 1948 (Turnbull, 2002). In 1956, George Town become the first municipality in the Malayan Federation under an elected local council. On 1 January 1957, it was granted the city status by Queen Elizabeth II as being the first city in Malayan Federation or Malaysia (Wong, 2013). The jurisdiction of George Town city was later expanded to cover the entire Penang Island in 2015.

After the formation of the Malaysian federal government, George Town' free port status was revoked in 1969 while the federal government focused on developing Kuala Lumpur and Port Klang. Consequently, George Town started to decline and Penang's overall economic development was affected due to brain drain when there was a flight of talent out of the state (Goh, 2014). The inner city of George Town began to hollow out when property rentals started to increase due to the repeal of the Rent Control Act in 2001. Besides, George Town's development was also affected by ineffective urban planning policies and poor traffic management (Jenkins, 2008). Eventually, the situation spurred NGOs and stakeholders to form strategic partnerships to conserve George Town's heritage and restore the city to its former glory. In 2008, together with Malacca, George Town was listed as a UNESCO World Heritage Site (WHS) for fulfilling three criteria of the Outstanding Universal Value (OUV), namely, criterion (ii) multicultural trading port; (iii) multicultural community, which refers to the living testimony of the multicultural heritage and tradition of Asia, and European colonial influences; and (iv) a unique architecture, culture and townscape multi-trading port (shophouses and townhouses) (UNESCO, 2008). The multicultural traditions and living practices by George Town's multi-ethnic residents offer rich tangible and intangible cultural heritage displayed through the city's architecture, gastronomy, art, heritage, religious practices and cultural festivals.

The post-UNESCO inscription scenario shows flourishing urban economy with vibrant tourism activities and emerging cultural and heritage-related businesses within George Town. Correspondingly, the local socio-economic activities are regenerated through a spin-off effect and the state government began to focus on upgrading the city's traffic flow, cultural and environmental aspects. The city has drawn in visitors and new residents, both locals and migrant communities, who stay, work, visit or engage in different social and cultural activities held within George Town. In recent years, new, trendy and imported cultures such as pop-up markets and hipster cafés are emerging within George Town World Heritage Site (GTWHS).

Before discussing George Town's situation in creative city-making, a brief review of Penang state is essential. Penang state is divided into Penang Island and Seberang Perai (formerly Province Wellesley). In 2019, the population of Penang Island and Seberang Perai were 817,200 and 970,000 respectively. In total, Penang state has an estimated population of 1,780,600, representing 5.4% of the national population (DoSM, 2020a). Penang is the third most densely populated Malaysian state, after Kuala Lumpur and Putrajaya (DoSM, 2019). As per 2019 data, a household in Penang earns a median monthly income of RM6169 and mean monthly income of RM7774. Overall, Penang registered 1.9% incidence of absolute poverty (DoSM, 2020b). Broadly, Penang state's economic performance strongly relies on services sector (contributing 50.6% of Penang's GDP in 2018) and manufacturing sector (contributing 43.3% of Penang's GDP in 2018) (DoSM, 2020a). However, the economic base in George Town inner city per se is predominantly on cultural resources and heritage endowments instead of manufacturing (Khoo et al., 2014).

Undeniably, culture-led urban regeneration is widely playing out in George Town, especially after the 2008 UNESCO inscription. The state and private sectors alike (especially small proprietors) are responding by harnessing George Town's cultural endowments especially within the world heritage site. While celebrating cultural heritage has been integral to George Town given the city's rich cultural diversity, the cultural turn was more deeply felt post-UNESCO inscription. State policy-makers started to invest in arts, cultural and creative initiatives as part of the city's urban regeneration strategy. This supports Curtis' findings (2009) that new economic opportunities and urban entrepreneurism that hinged on 'creative economy' and 'cultural economy' tend to emerge and flourish within world heritage sites. George Town is no exception. Nevertheless, the

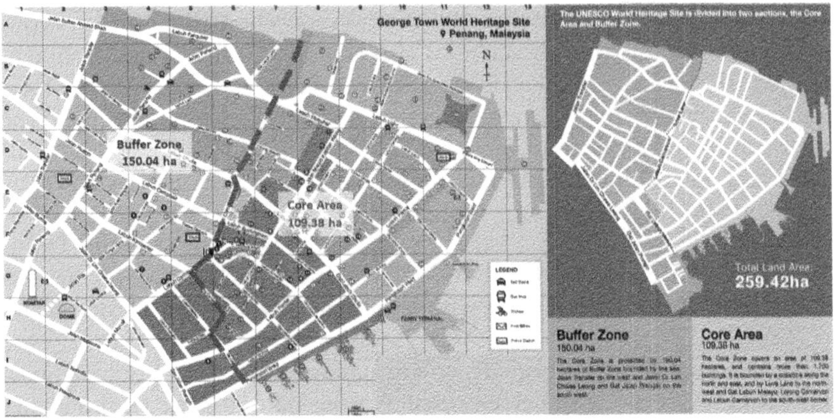

Fig. 6.1 Location of George Town World Heritage Site (Core Area & Buffer Zone). (Source: GTWHI (George Town World Heritage Incorporated), 2015)

route towards a culture-led creative city for George Town has upsides and downsides. While the city boasts a myriad of cultural and creative endowments that can be harnessed for inclusive local development, George Town in its current form has also succumbed to the downsides of indiscriminate heritage tourism and gentrification (Foo & Krishnapillai, 2018). Figure 6.1 shows the location of GTWHS.

6.2 Situation Analysis of Creative City Strategy for GTWHS

The following sections will discuss the viability of the creative city strategy for GTWHS.

6.2.1 George Town's Outstanding Universal Value (OUV)

George Town (GT) possesses rich and diverse cultural heritage assets that are manifested in both tangible and intangible forms. Specifically, these assets fulfilled three out of 10 criteria of UNESCO's OUV. Under Criterion (ii), GT is an exceptional representation of a multicultural trading town in East and Southeast Asia. It was established from the trade and exchanges of different cultures among the Malay, Chinese, Indian and the three consecutive European colonial powers. Each cultural group marked its cultural imprints within GT's cityscape, on the architecture, urban form and urban design. Next, Criterion (iii) recognises GT as a living testimony of the multicultural tangible and intangible heritage shaped under the Asian and European influences. These are portrayed through varieties of religious buildings of varying faiths, ethnic groups, diverse languages, beliefs and religious festivals, dances, costumes, art and music, gastronomy and daily life. Finally, Criterion (iv) acknowledges GT's eclecticism of culture and traditions which are reflected by the unique architecture, cultural forms and cityscape that are unparallel within East and South Asia. In particularly, GT preserved an exceptional range of shophouses and townhouses that evolved since the colonial periods (see Fig. 6.2). The GTWHS consists of a core zone (109.38 hectares) and a buffer zone (150.04 hectares) with a total of 5013 heritage buildings.

In terms of intangible cultural heritage, GTWHS has a legacy of traditional living cultures and historic enterprises. A total of 607 old traders

Fig. 6.2 Examples of heritage shophouses within GTWHS. (© Suet Leng Khoo (July, 2018))

and artisans, under 63 different classifications, strive to sustain their traditional trades, art and crafts that are inherited from generation to generation (GTWHI, 2012). They are key suppliers of George Town's intangible cultural heritage. However, many of them are finding it challenging to get a successor to sustain their sunset heritage trades (Khoo, 2020). This indicates that GT's intangible heritage is under threat of extinction.

The challenge to drive GT as an inclusive creative city is about the way to effectively conserve heritage while at the same time encourage creativity to flourish within tolerable parameters. It is not easy. Khoo's (2020) recent study highlighted the 'non-symbiotic co-existence of traditional artisans/heritage traders and modern artists/creative entrepreneurs' as GTWHS gentrifies after UNESCO's inscription. Current developments and conversion of heritage buildings for new purposes (e.g. modern cafés, bistros, boutique hotels, museums) (see Figs. 6.3 and 6.4) by local young entrepreneurs or foreign investors have significantly altered GTWHS's

Fig. 6.3 Modern eatery occupying two heritage shophouses at Beach Street. (© Suet Leng Khoo (July, 2018))

Fig. 6.4 A boutique hotel occupying shophouses at *Jalan Masjid Kapitan Keling*. (© Suet Leng Khoo (June, 2018))

original urbanscape. This challenged the original urban dwellers, especially the elderly tenants and old traders, to re-adapt themselves and to blend into the new living environment together with their new and younger neighbours. Heritage conservationists view these irrelevant new elements as a threat that may destroy the city's soul (Khoo, 2014). Furthermore, the increase in property prices in GTWHS has gradually evicted and displaced GTWHS's original residents which include historic traders and artisans who find it unaffordable to rent or own their old properties in GT. According to GTWHI's inventory, 20% of traditional traders, craftsmen and artisans have disappeared since 2012 (Teoh, 2018). This might threaten the heritage status due to the loss of intangible cultural heritage that reflects Criterion (iii) of the city's OUV.

6.2.2 *George Town and the City's Emerging Art Movement*

The cultural turn in GT, arguably, became distinctive after the UNESCO inscription. The state government actively invested in arts and culture towards promoting GT as an art and cultural hub alongside its heritage branding. Three major movements that contributed towards rejuvenating GT's vibrancy are (i) street art movement, (ii) George Town Festival (GTF) and (iii) vibrant art exhibitions and events. With the attempt to landmark GT's main streets, the state government initiated a street art project entitled 'Marking George Town' in 2009. A total of 52 story-telling caricatures and steel-rod sculptures were installed on the streets to raise public awareness of the local heritage. Inaugurated in 2010, George Town Festival (GTF) (a month-long arts festival) was held in conjunction with celebrating the UNESCO inscription. Ever since, arts festival and cultural events that include performances (e.g. theatre, music, dance, opera, film, etc.), exhibitions (visual arts), installations (public art) and cultural programmes were held with both local and international arts and cultural practitioners' participation. In 2012, as part of GTF events, the project 'Mirrors George Town' added in a collection of state-commissioned murals within GTWHS. The murals painted by Lithuanian street artist, Ernest Zacharevic, had gone viral in social media and captured global attention. Since then GT's street art became one of tourists' favourite Instagrammable attractions. Besides street art movements and GTF, the long existed fine arts sector that developed since the 1940s remains actively vibrant. Art exhibitions are consecutively held each month at galleries and exhibition venues around GT.

Under Penang State Executive Council (EXCO) for Tourism Development, Arts, Culture and Heritage (PETACH) and Penang Global Tourism (PGT) offices, the state government has constantly supported small- and large-scale art and cultural events held in GT each year. Besides government-supported events, privately initiated arts and cultural events collectively made GT eventful. However, creative and cultural endeavours in GT are heavily funded by the state with only pockets of funding from the private sectors. Nonetheless, these efforts collectively enhance local exposure and awareness towards arts and culture thus nurturing local consumption and demand for arts and culture. GT is increasingly transforming to become a popular art and cultural city within the region where the city has attracted an influx of visitors, art and cultural practitioners and creative talents. However, some challenges are encountered along the way.

Albeit the international hype and presence of art events, GTF's outreach and efforts to engage participation from different levels/classes of local society are questionable as highlighted here.

> It (GTF) does not seem to feature many local artists. Most of the art shows are exclusive, with international standards and cater to those who know how to appreciate or understand arts. Yes, it is well known internationally. Yet, still many locals, especially the local public do not know much about GTF. (Art curator-cum-artists promoter, interview, 2 March 2019)

According to GTF's director, Joe Sidek, who has organised the festival for nine years consecutively, the poor participation from local art practitioners is a cause of concern as he only received four local proposals out of all the proposals he had received from 15 countries in 2016. He highlighted that many locals hesitated to participate due to fear of not making the mark (Nambiar, 2016), thus, questioning the quality of creative talents in GT. While outreach of the festival is still an issue, another concern is the state government's move to cut down funding from RM4.5 million in 2017 and 2018 to RM3 million in 2019. The duration too has been shortened from one month to 16 days under a new event organiser (The Star, 2019). These raise concerns over the sustainability of the festival. Furthermore, it questions the state's commitment towards inclusive art and cultural diversity.

Besides the challenge to ensure equal accessibility to arts and culture, the emergence of new arts and cultural contents have brought up controversies and challenged local tolerance. Interview responses were ambivalent and revealed both acceptance and rejection towards new arts contents that emerged and flourished within George Town. The most controversial issue is over the new identity of George Town as an international street art capital (GT got listed in the Lonely Planet book of Street Art in 2017). There are advocates and critics on the existing street art movement in GT. On one hand, according to Chang's study (2018), key stakeholders in George Town's creative and cultural sector acknowledged street art as part and parcel of GT's local culture. Street or public artworks in GT has become GT's main attraction. By seeing more young people in GT, Hin Depot's proprietor (a creative hub in George Town) acknowledged the role of murals in transforming a heritage town (GT) into a vibrant creative city with an old backdrop (Chang, 2018).

Contrastively, objections were raised pertaining concerns of losing GT's local identities. The debate was rather intense during the first international street art festival (i.e. *Urban Xchange*) held in 2014. The intensity of the debate was observed in a public forum held during then (Tritto & Chang, 2019). In the same year, the state government set up the Public Art Review Panel (PARP) to regulate public art movement within GTWHS. Chang's study (2018) revealed the contradicting values between heritage conservation and freedom of creative artistic expressions that constantly constrain the development of local creativity. From an artist's viewpoint, the regulation on public art is rather biased. It limits the creativity of self-initiated artistic projects by the public for the public. However, a PARP's panellist insisted on the rationale of setting a limit to the burgeoning artworks on the city's walls in order to protect the city's heritage status.

Apart from Chang's (2018) findings, some interviewees expressed disagreement with the tourism-oriented commercialisation of art and cultural attractions in GTWHS, which may affect the cityscape's authenticity.

> The street art scene in GT has exploded, brought in new economic growth and tourism boom to the city. The trend has been accelerated rapidly, especially through social media. It has then swept away Malaysian culture swiftly. (Artist, interview, 19 August 2019)

> Armenian Street used to be more pleasant to walk. It is now too commercial and too tourism-oriented. It's always packed with tourists who are eagerly taking selfies with the murals. (Art-cum-craft shop retailer, interview, 10 March 2019)

Ernest Zacharevic, whose murals became the main crowd-puller in GT since 2012, has recently expressed his regret to see his iconic mural of *Children on a Bicycle* at Armenian Street as being the cause that transformed the heritage site into a 'circus' (see Figs. 6.5 and 6.6). He blamed his mural for the over-tourism in GTWHS. He even considered painting over his mural. He wrote this long emotional post on his Instagram (Today, 2019).

> *The street has not been the same as it used to when I first moved there, quiet heritage street with few local residents offering antiques or RM6 haircut on a ground floor of their family home which have been replaced with souvenir shops, restaurants, and all kind of insta friendly consumable concept stores to satisfy*

Fig. 6.5 Visitors taking pictures with the mural *Children on a Bicycle* at Armenian Street. (© Suet Leng Khoo (July, 2018))

> *ever-increasing traffic of holiday goers looking for 'authentic Penang experience' … Myself and many others blame my work for Armenian Street being a centre of the tourist route in Penang … I've been contemplating of simply painting over it in hopes to put an end to that circus…*

6.2.3 George Town's Creative Hubs and Creative Entrepreneurship

In the last decade, the emergence of creative hubs in GT somewhat indicated the adoption of the creative city concept in the city. By far, the most notable creative hub in GT is Hin Bus Depot Art Centre which started since 2014. Hin Depot is an exemplary private initiated art-led urban revitalisation project. The old depot has been adaptively reused and transformed into a creative cluster for creatives where it currently serves as a community art space in GT. The Depot's main functions are to facilitate

Fig. 6.6 A souvenir stall in front of a heritage shophouse at Armenian Street. (© Suet Leng Khoo (July, 2018))

the nurturing of art and creative talents as well as support collaboration and incubation of creative entrepreneurship. A small creative collective that works alongside local communities sustain the hub with active creative initiatives. The Depot, which expanded from one gallery space into a cluster with growing creative contents (see Figs. 6.7 and 6.8), has not only revitalised the old bus depot but also its neighbourhood. Gentrification took place with new developments and property investments that consequently boosted up the property prices around the vicinity. With an attempt to minimise the impact of displacement or drastic changes to the original neighbourhood, Hin Depot's operator-cum-proprietor acted mindfully and shared his views as quoted here.

> The land value has significantly increased. Although we encountered multiple inquiries to invest or to buy off ... or to rent the whole site for commercial business, yet, as *Penangites*, while we still can afford, we prefer

Fig. 6.7 The gallery cum multipurpose space at Hin Depot. (© Suet Leng Khoo (December, 2018))

> to remain it as a community place to promote local arts and creative entrepreneurs towards benefitting the local community. We consciously control ourselves not to be tempted by high rental despite we struggled with income in the beginning. Not easy … the old tenants used to pay a very cheap rental (RM500-RM600). First three years, we did not raise their rental. We gradually raised RM100 and now we managed to increase the old rental near to market rate (RM1800 – RM2000). Currently, two old tenants are still operating and servicing our community here. (Operator-cum-proprietor of Hin Depot, interview, 9 March 2019)

Hin Depot's initiative shows an example of how gentrification contributes towards revitalising the physical neighbourhood without neglecting its local community. The collective efforts in making Hin Depot a community's creative cluster in GT resonates the processes of 'civic boosterism' and 'post-industrial urban entrepreneurialism' (Harvey 1989) (cited in Karvelyte, 2018) but in a rather inclusive and sustainable way. The

Fig. 6.8 The Sunday Pop-up Market at Hin Depot. (© Suet Leng Khoo (December, 2018))

process was initiated and grew organically at the grassroots level towards benefitting local communities. Evidently, the creative strategy could be driven by a bottom-up force which is extremely crucial in steering the development and sustainability of a creative city.

Apart from art galleries, museums, creative marketplaces and street art, GT has witnessed the emergence of many hip and arty cafés, trendy eating places and concept shops within its vicinity. Some of these spaces frequently host art and cultural events. Besides art and culture, GT is also a hub for creative multimedia content (CMC). The state government had restored the renowned *Wisma Yeap Chor Ee* heritage building into a centre for incubation and development of creative multimedia and content entrepreneurship in Penang. The property was renamed as Penang Creative Animation Triggers (CAT) in 2014. The set-up of digital infra and infostructure at GT has attracted over 10 local CMC companies and a Shanghai-based animation studio to start up or expand their creative businesses in Penang. The growing creative contents within the city has drawn the influx of creative entrepreneurs and agglomeration of creative talent in GT. Although this helps to spin-off local businesses and complement the local tourism industry, their presence also poses a concern towards the survival of traditional traders and segregates GT's working-class communities. Currently, original urban dwellers strive to secure affordable housing within the gentrified neighbourhood. This is aligned with Khoo's recent study (2020) that highlighted social segregation, socio-economic inequality and unaffordable housing as key challenges towards making GTWHS a socially inclusive creative city.

6.2.4 Fragmented and Non-cohesive Creative Ecosystem

Despite GT having strong creative and cultural initiatives, the creative ecosystem is rather disconnected and disorganised due to fragmented endeavours. There are diverse creative individuals or groups from different fields delineated across ethnicities and cultural practices driving different creative sectors independently. Each tends to cluster among their own group of communities. The disconnection is not only happening between different creative sectors but also within the value-chain of the same creative sector such as the arts ecosystem. Within the arts ecosystem itself, artists, art groups and art marketers work individually and gather in different groupings with different focus areas/aspirations in different art fields. Recently, Khoo's (2020) study revealed that fine artists tend to be more 'individualistic, soloistic and egoistic' as they strive to showcase their own identity and personality. These discoveries challenge the development of a cohesive art ecosystem. Despite collaborations happening occasionally, it is merely limited among the same active circles. The impacts hardly reach out effectively to all levels. Therefore, inter and intra-sectoral collaborations are important to consolidate creative energies towards generating impactful shared values among all. Nonetheless, without a strategic masterplan to drive the development of the creative economy as a whole, GT's creative sectors seem to be operating non-cohesively without a clear and proper direction.

Overall, GT's vibrant creative vibes mostly come from the art and cultural sectors. As a state body, PETACH plays an important role to promote Penang's arts and culture. However, the efficiency and effectiveness of PETACH in performing its assigned roles are rather doubtful given that it is assuming the responsibilities to lead and drive all four sectors (i.e.

tourism, art, culture and heritage) at the same time. Each of these fours sectors covers wide and diverse scopes independently. Each operates under a different ecosystem, involves different industrial settings and diverse nature of production or consumption. Each sector is driven by different groups of stakeholders who are encountering different needs and challenges. Therefore, relying on one single department to plan, decide and drive the four different sectors in Penang is impractical, either in terms of finance or policy direction for the development of each sector (tourism, art, culture or heritage). Currently, both PETACH and PGT are eagerly promoting Penang or GT as an eventful city through arts, culture and heritage to drive local tourism. The state's tourism-oriented initiatives, however, have raised concerns among local practitioners on the negative consequences due to commodification of culture, arts and heritage. The following interview responses show concerns of losing the actual value of art, culture or heritage when these elements are being over-packaged mainly for tourists' consumption.

> The events in GT are mostly meant for tourists. I hardly go to town while a festival or event is happening. Like the Penang International Food Festival, the *Char Koay Teow* or *Chendol* is expensive and not as authentic as those I can find at the local hawker stalls. (Urban dweller, interview, 1 March 2020)

> Every politician has a political agenda. Without much relevant knowledge or understanding of art, it is really hard to expect the department (PETACH) to know how to support the art sector practically. (Art academy management personnel, interview, 15 June 2020)

> As the project funder, they (PETACH) decide the types of art events and the subject or contents of art to be presented to the audiences. Yet, their selection consideration is always limited to commonly accepted subjects that could fast capture the public's attention. They want to see fast results and tend to evaluate the impact of an event based on the popularity by counting the followers at social media. They support art initiatives not for art's sake but instead more towards the spin-off effect on local tourism. (Freelance art practitioner, interview, 3 March 2020)

After the recent state's restructuring, PETACH now carries its new acronym as 'PETACE' (Penang State EXCO for Tourism and Creative Economy) where 'arts, culture and heritage' are subsumed under the umbrella of 'Creative Economy' (Buletin Mutiara, 2020). This is rather inconsistent with Malaysia's National Creative Industry Policy definition of creative industries because PETACE only focuses on two sectors instead of three (i.e. multimedia, cultural arts, and cultural heritage). Furthermore, the creative economy could involve much wider and diverse sectors beyond arts, culture and heritage. Thus, it is unwise to conveniently subsume 'arts, culture and heritage' under the creative economy without first devising a standard operational definition for the city.

6.3 The Nexus Between State, Market and Civil Society

In GT, there are indications of the creative city concept being locally adopted due to the cultural turn that happened in the city post-UNESCO listing. The state government started to invest in arts and cultural initiatives to create awareness and promote local appreciation towards conservation of GT's cultural heritage as a world heritage site. The state-driven approach had consequently encouraged the clustering of art and cultural practitioners (local as well as international) in GT. Subsequently, the state's efforts have caused a growth of creative clusters with a mix of different urban creative entrepreneurship. The international cross-cultural collaborative initiatives (during GTF) and the growing contemporary creative contents within the historic city have collectively made GT unique and welcoming especially to tourists. The creative ambience within the city has also inspired more creative initiatives from the grassroots illustrated through active participation among local creatives though more cohesive efforts are welcomed among them. The bottom-up creative initiatives complement the state government's efforts in making GT a creative city, leveraging its art and cultural endowments.

Despite having active support from Penang's government, there is no specific masterplan to strategically drive the development of creative sectors in diverse fields. Nevertheless, the Penang 2030 Vision has included the development of creative industries and niche business services as one of its overarching strategic initiatives towards upgrading the local economy and household income (Penang2030, 2019). It emphasises the development of creative talents, products and services. Overall, it envisions Penang's creative and cultural economy will contribute 5% (instead of less than 1% in 2016) to the national GDP by 2030. Also, it targets to double the number of community-based arts/cultural productions by investing in museums, galleries network and developing a culture and arts masterplan. No doubt, the state plays a strong interventionist role to facilitate Penang's creative ecosystem. Nonetheless, these planned strategies, arguably, are catalytic top-down drivers which are rather supply-driven in nature. A good example of state-driven intervention in the art sector is the long-awaited plan to construct Penang Art District (PAD). The initial plan was announced in 2015 with RM70 million investment from Penang Development Corporation (PDC) (the state government's development agency). In 2017, the project was relocated to a new site and redesigned as a RM30 million development. Yet, the physical development of PAD has not materialised until today. The long-delayed project and the trimmed development budget somehow raise doubts about the state government's agenda. Hopefully, it will not end up as a 'blank cheque' from the government.

Based on GT's situation, relying on the state's efforts and the fragmented initiatives from the privates and the local creatives may not be sustainable in the long run. The non-cohesive creative ecosystem challenges the development of an inclusive and sustainable market of production and consumption in GT. Effective cross-sectoral stakeholders' partnership plays a crucial role as the way forward.

References

Buletin Mutiara. (2020, October 1). *PETACE to Play Pivotal Role in Local Economy, Says Exco Man*. Retrieved from: https://www.buletinmutiara.com/petace-to-play-pivotal-role-in-local-economy-says-exco-man/. Accessed 15 Oct 2020.

Chang, N. (2018, February). "Placemaking" Penang's Public Art. *Penang Monthly*, Issue February, 2018. Penang Institute.

Curtis, T. (2009, October 8–9). *Making the Most of World Heritage: Promoting Economic Growth and Heritage Conservation*. Penang International Conference on Sustainable Cultural Development: The Economics of Heritage Revitalization. George Town, Penang, Malaysia.

DoSM (Department of Statistics Malaysia). (2019, October). *Malaysia Statistical Handbook 2019*. Department of Statistics, Malaysia.

DoSM (Department of Statistics Malaysia). (2020a, May). *Poket Stats Negeri Pulau Pinang ST1 2020*. Department of Statistics, Malaysia.

DoSM (Department of Statistics Malaysia). (2020b, July). *Household Income & Basic Amenities Survey Report by State and Administrative District – Pulau Pinang 2019*. Department of Statistics, Malaysia.

Encyclopaedia Britannica. (n.d.). *George Town, Malaysia*. Encyclopaedia Britannica online. Retrieved from: https://www.britannica.com/place/George-Town-Malaysia. Accessed 28 Apr 2020.

Foo, R., & Krishnapillai, G. (2018). Preserving the Intangible Living Heritage in the George Town World Heritage Site, Malaysia. *Journal of Heritage Tourism, 14*(2), 1–13.

Goh, D. P. S. (2014). Between History and Heritage: Post-Colonialism, Globalisation, and the Remaking of Malacca, Penang and Singapore. *TRaNS: Trans-Regional and -National Studies of Southeast Asia*, *2*(1), 79–101.

GTWHI (George Town World Heritage Incorporated). (2012). *Directory of Traditional Trades and Occupations in George Town World Heritage Site 2012 (by Trades)*. GTWHI.

GTWHI (George Town World Heritage Incorporated). (2015). *Historic City of George Town – World Heritage Site*. Map. Retrieved from: http://gtwhi.com.my/wp-content/uploads/2019/05/GTWHI_MAP-20150226-web.pdf. Accessed 20 Nov 2020.

Jenkins, G. (2008). *Contested Space: Cultural Heritage and Identity Reconstructions: Conservation Strategies Within a Developing Asian City*. LIT Verlag. ISBN 9783825813666.

Karvelyte, K. (2018). Shifting Meanings in Changing Contexts: The Role of the Creative City in Shanghai, Hong Kong and Taipei. *International Journal of Cultural Policy*. https://doi.org/10.1080/10286632.2018.1479748

Khoo, S. L. (2020) *Towards a Socially-Inclusive Creative City: What George Town World Heritage Site Can Learn from Japan?* Technical Report. Malaysia, Penang: Universiti Sains Malaysia.

Khoo, S. L., Badarulzaman, N., Samat, N., & Dawood, S. R. S. (2014). Capitalising on Urban Cultural Resources for Creative City Development: A Conceptual Review and the Way Forward for Malaysia's George Town. *GEOGRAFIA Online Malaysian Journal of Society and Space, 10*(5), 20–29.

Khoo, S. N. (2014). George Town: The Discreet Charm of Rejuvenated Heritage. In Asian Cities: Cultural Heritage and the Interplay Between Nation Building and Internationalism. *Australia ICOMOS Historic Environment, 26*(3), 54–71.

Langdon, M. (2014). *George Town's Historic Commercial and Civic Precincts*. George Town World Heritage Incorporated (GTWHI).

Nambiar, P. (2016, December 19). *More Local Flavour Needed for George Town Festival. Free Malaysia Today (FMT)*. Retrieved from: https://www.freemalaysiatoday.com/category/nation/2016/12/19/more-local-flavour-needed-for-george-town-festival/. Accessed 8 Apr 2020.

Penang2030. (2019). *Penang2030: A Family-Focused Green and Smart State that Inspires the Nation*. Retrieved from: https://www.penang2030.com/files/The%20Penang2030%20Guide_First%20Edition%202019_eBook_.pdf. Accessed 30 Apr 2020.

Teoh, S. (2018, January 16). *UNESCO Listing Both a Boon and a Bane for George Town*. The Straits Times. Retrieved from: https://www.straitstimes.com/asia/se-asia/unesco-listing-both-a-boon-and-bane-for-george-town. Accessed 28 Apr 2020.

The Star. (2019, June 4). *George Town Festival Shortened with Smaller Budget*. The Star Online. Retrieved from: https://www.thestar.com.my/news/nation/2019/06/04/george-town-festival-shortened-with-smaller-budget. Accessed 28 Apr 2020.

Today. (2019, July 3). *Ernest Zacharevic Blames His Art for Turning Once-Quiet Heritage Street in Penang into 'Circus'*. Today Online. Retrieved from: https://www.todayonline.com/world/ernest-zacharevic-blames-his-art-turning-once-quiet-heritage-street-penang-circus. Accessed 28 Apr 2020.

Tritto, A., & Chang, N. S. F. (2019). Destination Management Through Stakeholders' Collaborations. In N. Kozak & M. Kozak (Eds.), *Tourist Destination Management: Instruments, Products and Case Studies*. Springer International Publishing.

Turnbull, C. M. (2002, April 18–21). *Penang's Changing Role in the Straits Settlements, 1826–1946*. The Penang Story – International Conference 2002, The City Bayview Hotel, Penang, Malaysia.

UNESCO. (2008). *Melaka and George Town, Historic Cities of the Straits of Malacca*. Retrieved from: http://whc.unesco.org/en/list/1223. Accessed 28 Apr 2020.

Wong, C. W. (2013, October 5). *George Town's First Mayor Ramanathan Was a Fiery Man, Politician, Educationist and Unionist*. The Star Online. Retrieved from: https://www.thestar.com.my/news/community/2013/10/05/george-towns-first-mayor-a-fiery-man. Accessed 28 Apr 2020.

CHAPTER 7

Ipoh – From Tin Mining City to Creative City

Abstract This chapter discusses the viability of creative city as an urban regeneration strategy for Ipoh city, a former tin mining city. It explains the local council's adoption of a conservation-centric strategy towards regenerating Ipoh city centre through creative place-making. Besides analysing the current scenario in Ipoh, this chapter also discusses the nexus of the state, private sector and civil society as Ipoh embarks on a creative city pathway.

Keywords Creative city • Urban regeneration • Ipoh • Perak

7.1 Ipoh – Capital City of Perak

Ipoh city is located on Kinta River at west Peninsular Malaysia and is the capital of Perak state. It is a two-hour drive from KL (at the South) and from Penang (at the North) through the North-South Expressway. Ipoh city is a popular transit city for those traveling from KL to Penang or vice versa. Ipoh is the most important station on the KL-Butterworth railway that connects to cities in the North and is the main connecting city along the highways across the valley (Encyclopaedia Britannica, n.d.).

Ipoh was founded in 1500s as settlements for Sumatran Malay indigenous people along the Kinta River banks (Wee et al., 2014). Ipoh's history started with the historic centre of Ipoh old town, which is separated from

S. L. Khoo, N. S. F. Chang, *Creative City as an Urban Development Strategy*, https://doi.org/10.1007/978-981-16-1291-6_7

its new town by Kinta River. It flourished as a town during the tin rush in 1870s when British ruled Perak around 1877. Local royals and Chinese businessmen started to build shophouses and set up businesses around Ipoh old town. In 1884, it became a busy town with wide straight roads connecting areas around Kinta River. According to Khoo and Lubis (2005), by 1890, Ipoh had 340 houses that catered to a population of about 4000. After an official restructuring in 1892 to restore the damages caused by a great fire, Ipoh was rebuilt and became the largest town in Perak and the second largest in the Federated Malay States in 1895. British tin-mining companies transferred operations from Taiping to the rich Kinta ores where immigrant Chinese were brought in to work (Encyclopaedia Britannica, n.d.). Ipoh was then facilitated with improved roads system linking Ipoh to other places around Kinta Valley and equipped with other basic infrastructure services and community spaces. The old town continued to swell until 1907 when a new township was commissioned (Khoo & Lubis, 2005). Since then until 1914, Ipoh town administrative area was expanded across the Kinta River where Ipoh new town began to develop. More houses and buildings were built in Ipoh new town, including train station, hotel, cinema, food market, town hall and post offices.

Due to the flourishing tin mining and rubber production industries, Ipoh grew rapidly throughout the 1920s and 1930s until it was invaded by the Japanese in 1941. During the Japanese occupation, Ipoh replaced Taiping to become the capital of Perak. With rapid development and growing population, Ipoh gained its Municipal status in 1962 and later its City status in 1988. Today, it is governed under Ipoh City Council (ICC), headed by a Mayor. The continuous development of Ipoh new town and other new emerging townships within Kinta District have eventually dispersed the concentration of administrative, economic, commercial as well as city services activities beyond Ipoh old town. Consequently, Ipoh old town underwent the same process as other traditional core, including depopulation, ex-urbanisation and degeneration (Clark, 1989; Steinberg, 1996). Subsequently, many urban functions were gradually decentralised to the new town and other suburban areas.

According to Ipoh Local Plan 2020, Ipoh City Council aims to develop Ipoh by focusing on urbanisation & sustainable resources; integrated economic development; efficient communication & accessibility; and to build prosperous communities. The local council strategises to develop Ipoh city centre into an important hub for businesses, services, main industries and key functions (ICC & JPBD, 2012). In 2014, ICC drafted Special Area Plan (SAP) of Ipoh Town-Tin Heritage City 2020 to specifically drive regeneration initiatives within both the old and new towns. Figure 7.1 shows the Core Zone (79.70 ha.) and Buffer Zone (175.50 ha.) for Ipoh Tin Heritage City (ITHC).

The National Physical Plan marked ITHC as a sub-regional growth conurbation centre, which links up Penang and KL's conurbations. Locally, ITHC plays its role as the state regional centre and the Central Business District (CBD) of Ipoh. Besides, it was listed as one of the 10 international tourism destination within SUPERI (Sungai Perak Indah) Tourism Zone (ICC, 2014).

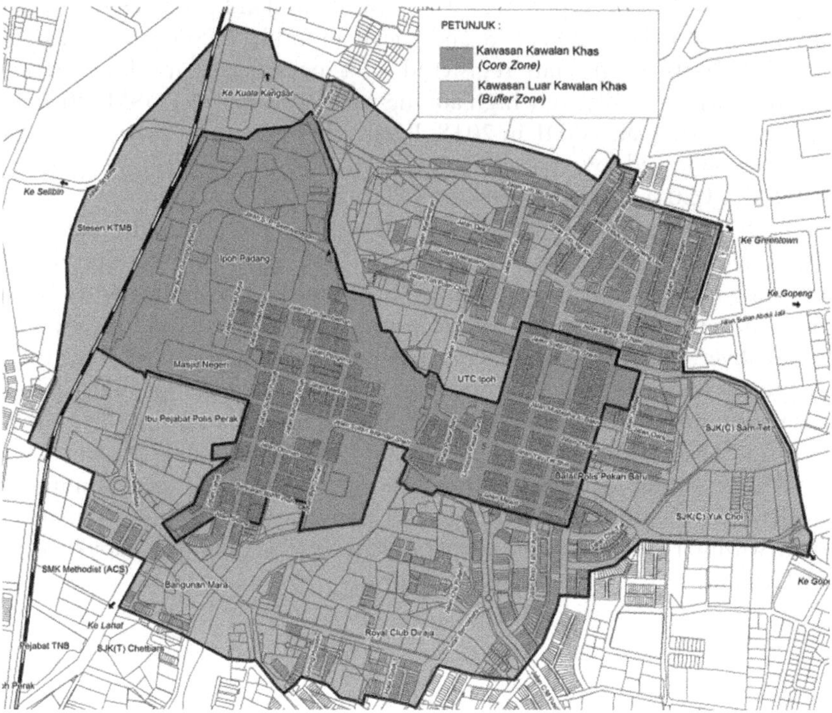

Fig. 7.1 Location of Ipoh Tin Heritage City (Core Zone & Buffer Zone). (Source: Special Area Plan (SAP) of Ipoh Town-Tin Heritage City 2020, ICC, 2014)

Broadly, the SAP adopted a conservation-centric regeneration strategy, leveraging Ipoh's local heritage and cultural identities. It aims to transform Ipoh town into a liveable national heritage city of tin. The plan aspires towards achieving economic, social, cultural and environmental sustainability through government-led or government-private collaborative initiatives. The emphasis is on restoring the city's former glory; enhancing creative place-making and city branding; promoting a sense of place and pride among locals by creating memorable experiences; and developing competitive destinations for heritage tourism within ITHC towards supporting Perak's development in 2030 (ICC, 2014).

As an overview of Perak, the state registered a population of 2,509,500 in 2019, which represents 7.7% of national population (DoSM, 2020a). If compared to KL, Penang and Johor, Perak has the lowest population density (DoSM, 2019). As per 2019 data, a household in Perak earns a median monthly income of RM4273 and mean monthly income of RM5645 (DoSM, 2020b). The state registered 7.3% incidence of absolute poverty in 2019, which is relatively high among Malaysian states (DoSM, 2020b). According to Perak's GDP in 2018, Perak relies strongly on services sector (62.9%) followed by manufacturing (18.1%) and agriculture (15.1%) (DoSM, 2020a).

Both Ipoh Local Plan 2020 and SAP of Ipoh Town-Tin Heritage City 2020 steer the local council's revitalisation strategies at ITHC. The efforts started with conservation of heritage buildings and monuments; restoration and enhancement of heritage tourism attractions; beautification of Kinta River; creative place-making and pedestrianisation of heritage trails at the old and new towns. Seemingly, ITHC is undergoing similar transformation milestones as what George Town and Kuala Lumpur had experienced. The cultural turn slowly became significant in ITHC when the state government, private sector, civil society groups and local communities started to initiate and engage in creative art and cultural initiatives held within ITHC at different periods. The first initiative happened in 2012 and it was a private-driven initiative.

7.2 Situation Analysis of Creative City Strategy for Ipoh Tin Heritage City (ITHC)

The following sections discuss the situation of creative city adoption in ITHC.

7.2.1 *Cultural Heritage Significance of ITHC*

Both the old and new towns located within ITHC preserve Ipoh's tangible and intangible cultural heritage. A total of 1540 heritage buildings and monuments remain intact, in which 14 units were gazetted under the National Heritage Act 2005 and Ipoh Local Plan 2020 respectively. The architecture styles of the heritage buildings portray the influences that mark a footprint in this tin city (ICC, 2014). The legacies of traditional living cultures include old trades, traditional artisans, food and cultural practices that collectively build up the cultural and creative endowments within the city.

Despite having structured initiatives in the SAP of Ipoh Town-Tin Heritage City 2020 to conserve, upgrade, promote and improve accessibility to Ipoh's tangible and intangible heritage assets, a few concerned Ipoh residents criticised the plan. They disagreed with the plan for emphasising on place-making the city as a heritage city, mainly for monetary returns through tourism. Instead, they urged the government to look into the communities' benefits. They questioned the government's approach in theming the city artificially into one single identity without considering the complexities, multiplicities and identities of the past, present and future generations of Ipoh people (Lee et al., 2014). This casts a doubt whether genuine public participation and bottom-up communication had been involved during the formulation of the SAP. Without adequate public participation, it is unlikely to achieve a socially inclusive creative city.

Although there is dissatisfaction with ineffective public participation in the planning stage, however, active local public engagement was observed during the execution phase. As a planned project to preserve and develop the heritage sites at *Jalan Dato' Maharajalela*, *Jalan Panglima*, *Lorong Panglima* (Concubine Lane) and *Jalan Bijih Timah*, the local council supported the private place-making initiative along Concubine Lane (a 120 years old lane) in 2014 (The Star, 2014; ICC, 2014). While the local council was upgrading the surrounding with soft and hard landscape elements, the property proprietors restored and renovated the dilapidated and vacant heritage shophouses for new occupancies. Under a public-private partnership, Concubine Lane became a popular pedestrian and tourist-friendly heritage-cum- cultural walk area, akin to Armenian Street in GT or Jonker Street in Malacca (see Fig. 7.2). Visitors can find souvenirs, art & crafts, local food and delicacies along the stalls at the front of the restored shophouses. Only a few restaurants retained their old business

Fig. 7.2 The heritage-cum-cultural walk at Concubine Lane. (© Nicole Shu Fun Chang (August, 2019))

within the newly transformed site (see Fig. 7.3). Subsequently, restoration and upgrading initiatives were observed at Hale Lane and Market Lane with murals, street deco and landscaping. The three adjacent lanes which are connected due to their interrelated history, are now revitalised with introduction of new contents that attract droves of visitors. A field visit revealed some inauthentic elements and irrelevant contents on sale that do not represent ITHC's identity but mainly target the tourism market. This raised concerns on losing the city's authenticity and identity as a result of over-tourism amidst unmindful planning and management.

The increasing foot traffic within ITHC has encouraged private initiatives to restore heritage buildings into private museums that showcase local folklores. For example, *Ho Yan Hor* Museum, *Han Chin Pet Soo* and 22 Hale Street (see Figs. 7.4, 7.5 and 7.6).

According to the curator of *Ho Yan Hor* Museum, the museum received about 100,000 local and international visitors annually (Bunyan, 2018).

Fig. 7.3 An old restaurant sustains its business at Concubine Lane. (© Nicole Shu Fun Chang (August, 2019))

Yet, 22 Hale Street which started its museum-cum-gallery since January 2019 is still struggling to attract visitors.

> We started with admission fee of RM15 and RM10. It did not work. So, we lowered it to RM10 and RM5, half price. Yet, people are still not willing to pay. (Curator of a museum, interview, 21 September 2019)

This indicates the inconsistent or yet to be established consumption market for heritage or cultural contents in ITHC. A local visitor pointed out that the locals are mostly interested in food instead of history or culture.

> *Ipohlites*, especially those back from outstation, mostly come for the food that they are longing for. They are willing to wait in long queues at traditional coffee shops instead of visit the museums. (Local visitor at ITHC, interview, 12 October 2019)

Fig. 7.4 *Ho Yan Hor* Museum displays the history of a local health tea brand. (© Nicole Shu Fun Chang (August, 2019))

Fig. 7.5 *Han Chin Pet Soo* – the *Hakka* Miners' Club & Museum. (© Nicole Shu Fun Chang (August, 2019))

Although the traditional coffee shops or eateries still run their family-inherited businesses well, a key informant is concerned about business sustainability due to absence of successors.

> Many operators are old. Their younger generations have mostly moved to KL, Penang or overseas for better job opportunities. It is truly not easy to get the next generation to commit to succeed the old coffeeshop business here. (Officer at Perak State Executive Councillor Office (EXCO) of Tourism, Arts and Culture, interview, 16 August, 2019)

Despite Ipoh's rich cultural heritage significance, limited local appreciation and lack of sense of pride of the community own's history and cultural heritage may challenge the preservation of the city's identity.

Fig. 7.6 22 Hale Street – a heritage museum-cum-gallery. (© Nicole Shu Fun Chang (September, 2019))

7.2.2 Growing Creative Contents within ITHC

The first creative content that emerged within ITHC was initiated by a few Ipoh natives in late 2012. They attempted to revive the deteriorating old town through real estate redevelopment projects and *Sekeping Kong Heng* at *Jalan Bandar Timah* was their success story. It is one of the series of *Sekeping* creative adaptive redevelopment projects designed by an Ipoh-born architect who is based in KL. Besides Ipoh, Sekeping projects are seen in Penang, Selangor and KL. Sekeping Kong Heng has successfully restored and adaptively reused the old building of local famous *Kong Heng Kopitiam* by transforming the first floor into hotel rooms with original and rustic architecture (Tunyaporn, 2018). Its adjacent warehouse space was turned into a big chain restaurant (see Fig. 7.7). A few old shophouses nearby were refurbished as spaces for local start-ups. The open courtyard that links up all entities became an artisan market for local creative entrepreneurs to start their art, craft and cultural stalls/shops (see

Fig. 7.7 The old warehouse is now a big chain restaurant at *Sekeping Kong Heng*. (© Nicole Shu Fun Chang (August, 2019))

Fig. 7.8 and 7.9). Sekeping Kong Heng emerged as the first creative cluster where creatives started to hang out for creative consumption and inspiration. It soon recorded active foot traffic, especially among the younger generations and creatives. As in many other creative clusters, urban entrepreneurialism gradually took place organically along the process. Local art and cultural practitioners started to initiate and engaged themselves in art or cultural-led place-making initiatives.

After seeing how Ernest Zacharevic's murals become the main crowd-puller in George Town, Ipoh's homegrown coffee chain, Old Town White Coffee commissioned Ernest to paint for a project, namely, 'Art of Old Town' in 2014 (Loh, 2014). Ernest painted seven murals depicting local lifestyles to pay tribute to Ipoh Old Town as the origin of white coffee (see Figs. 7.10 and 7.11). The completed murals were then officiated and endorsed by the local council. Immediately after the completion, the murals became Ipoh's instant attractions. The streets with Ernest's murals

Fig. 7.8 Pop-up art and craft stalls at *Kong Heng Square* Artisan Market. (© Nicole Shu Fun Chang (August, 2019))

started to witness influx of local and international selfie-takers or Instagrammers. The mural trail was then rapidly extended with more other murals created by local artists within the old and new towns. Unlike George Town's street art movement, Ipoh street art scene seems much more welcomed by the locals. It is possibly due to the themes of artworks that mostly revolve around local cultures which are less controversial.

In 2016, Ipoh city got listed as the sixth most attractive Asian destination to visit by Lonely Planet due to the flourishing hipster cafes around both the old and new towns (ICC, 2016). Additionally, Ipoh old town is popularly known as a food haven. Ipoh offers famous local delicacies that keep attracting returning visitors and Ipoh natives to purposely travel back for a food trip. The gastronomy, art and cultural vibrancy alongside its preserved heritage assets, collectively set a backdrop for creative city-making.

Fig. 7.9 A local creative start-up at the open courtyard of *Sekeping Kong Heng*. (© Nicole Shu Fun Chang (August, 2019))

ITHC's creative scene continues to flourish when PORT (People of Remarkable Talents – a state-owned art agency established since 2011), Tourism Perak (the state tourism agency) and some local art and cultural groups actively curate art and cultural events held at ITHC. Solely in 2019, multiple art festivals and events were held one after another to celebrate local creativity, arts and cultural identities. Although a regeneration impact was observed in ITHC, it mainly comes from the economic spin-off that benefits the local tourism-related sectors. It has yet to successfully generate direct income or trickle-down benefits to the creatives and the practitioners. As mentioned by the following key informant, the actual consumption and demand for local creativity are vital.

> The state and local government are ready to drive the creative city strategy. My main worry is the local acceptance, especially the local appreciation on creativity. Whether this creative market can be turned into commercial

Fig. 7.10 A signboard depicts the location of Ernest's murals at Ipoh old town. (© Nicole Shu Fun Chang (August, 2019))

> industries is the key to make it sustainable…so that artists and creative practitioners can survive… and those migrated local talents can come back. I always ask about the transaction, either in an art exhibition or our pop-up bazaar. That is not about how many visitors or how happening the event is. To survive, artists need sustainable income. (Perak State EXCO for Tourism, Arts and Culture, interview, 28 September 2019)

7.2.3 Underdeveloped and Underpromoted Creative Ecosystem

Despite the fact that the state and Ipoh City Council are keen in creative place-making (through arts and cultural initiatives in ITHC), there is no specific masterplan to strategise the development of local creative industries. The city lacks conducive creative infrastructure, both hard and soft. Besides a few private event spaces located at the main heritage site (near Concubine Lane), the ITHC has no public venues for art and creative activities, such as public galleries, performing centres and activity halls.

Fig. 7.11 Ernest's mural on a wall near *Jalan Panglima*. (© Nicole Shu Fun Chang (August, 2019))

Without the supportive hardware, Ipoh or Perak has encountered challenges to retain its creative talents, which is the software of a creative ecosystem. Many local creative talents or professionals opt to pursue their creative careers elsewhere instead of Ipoh. A respondent highlighted this and suggested a more practical approach based on the current situation.

> We have talents but mostly exported to Singapore, Penang, KL. Ipoh people are very pragmatic. They wanted to see fast return. Once they can't find the avenue to perform, they move out to try other options. At this stage of its development, I think what Ipoh creative sector needs is simply place-making and focusing on education and exposure towards creating local awareness and appreciation of creativity. (Curator of a museum, interview, 21 September, 2019)

According to PORT, diverse art talents are scattered around Perak. They are mostly underpromoted locally due to lack of baseline data and documentation.

> There are talents in Perak, especially in music. Ipoh was known for production of rock band music. Visual art in Perak was active once in 90s. We do have many artists, active art groups or collectives, and hidden talents in Perak. Yet, they are not linked up. We need to feature them. We need to work on a directory for art in Perak, compiling details of the local practitioners. Without such archive, many artists are not well recognised. They come and go. Some may have migrated to other states...we are losing our local talents. (General Manager of PORT, interview, 7 September, 2019)

Although PORT plays a role in creating platforms for creative talents towards making art more accessible in Ipoh, its focus seems limited to certain art groups or sectors only. In two separate interviews, the key informants pointed out the limited outreach of PORT's programmes in linking up the diverse groups of local art practitioners. Both highlighted the fragmented and incohesive initiatives by different groups.

> PORT is very Malay-focused. Their activities are very Malay-oriented. As a Chinese to participate in their event, I wouldn't feel very inclusive or easy to fit in. Like the rock-n-roll scene which Chinese are not really engaged in ... There are ways to open up this kind of scene to engage other races but it doesn't feel like the events are designed in that direction. Something needs to be done ... everyone here is doing their own thing now. Contents here are very much segregated by races. PORT is doing their Malay-focused programs while the Tin Alley group is very Chinese-oriented. And sadly, no content seems to be coming from Indians. (Curator of a museum, interview, 21 September, 2019)

> There are a few different active groups in Ipoh. Each has its own focus in arts or culture and very rarely see them collaborating. PORT's events usually concentrate on Malay culture or Malay artists. The state is trying to lead PORT to expand its scopes, to involve multi-ethnic practitioners and to link up the local groups. (Officer at Perak State EXCO Office of Tourism, Arts and Culture, interview, 16 August, 2019)

Apart from inadequate infrastructure and incohesive efforts that challenge the production of creativity in ITHC, limited local awareness and appreciation also affect creative consumption.

> The main problem is that locals do not see the value of their own culture and identity. Besides food, many don't know what else Ipoh could offer. The people here tend to have a shallow understanding of art and culture. Many see 'art' as a waste of time. (Founder of a cultural-cum-creative space, interview, 20 November 2019)

7.2.4 *Ipoh City as an Emerging Sub-Regional Creative Hub*

As previously mentioned, ITHC is identified as a sub-regional growth conurbation centre, which connects KL and Penang's conurbations. Thus, it facilitates inter-state dependency and collaboration in urban development towards achieving a regional shared vision or direction. In the creative city-making context, it probably could transcend beyond local or state level to a regional level. Regional partnerships in creative strategy could eventually drive a creative region, linking (potential) KL-Ipoh-Penang creative cities or creative ecosystems towards supporting regional growth. The mega creative ecosystem could be co-driven by the linked-up clusters that connect the creative workforce and industry players across the three states. An effective cross-state collaboration/partnership will develop a productive creative ecosystem in the country. Through exchange of ideas or experiences, co-creations, co-organisation or co-management of creative projects, it generates shared values and trickle-down benefits for more and bigger groups of communities, instead of limited to the same circles of creative talents in one city. Hopefully, with that attempt, there will be more winners instead of losers along the process of creative city-making in Malaysia.

Some cross-state collaborations were observed in Ipoh. The first example is during the Art Ipoh 2019. It was Ipoh's first-ever major art exhibition organised by Tin Alley Ipoh, a cultural-cum-creative space in ITHC. The exhibition showcased 36 installation art pieces created by 18 artists-in-residence (8 Ipoh artists and 10 artists from other states) at a row of dilapidated old shophouses along *Lorong Bijih Timah*. The whole project was directed by an experienced art practitioner from KL. Another example is during Ipoh International Art Festival (IIAF) 2019. IIAF was a week-long festival co-organised by PORT and Tourism Perak to make Ipoh a major city with its own signature annual art festival. The said festival received curatorial and programming support from National Art Gallery & partners and individual curators from other states. According to the organiser, the festival featured more than 100 art practitioners from different fields of arts presented by nine curating teams who originated from different states and some from South-East Asia. During both the Art Ipoh 2019 and IIAF 2019, ITHC received an influx of art lovers and art practitioners from other states who purposely came to visit and participate in the programmes. This indicates an expanded outreach of mobility and consumption of creativity through cross-state creative collaborations.

7.2.5 *Brain Drain*

Most of the working age cohort are keen to explore better career options in bigger Malaysian cities. Many Ipoh-born talents have ended up settling down in major cities like KL, Penang or Johor Bahru. According to a key informant, the cross-state migration happened since long ago. He highlighted brain drain as the main factor that caused traditional old trades in ITHC to gradually disappear.

> Since long ago, parents kept encouraging their younger generation to seek for better jobs outside of Ipoh. Some even encouraged their children to settle down elsewhere. So, at last, who is going to succeed their old businesses here (Ipoh old town). (Founder of a cultural-cum-creative space, interview, 20 November 2019)

As highlighted by another respondent, this brain drain scenario is also part of the contributing factor that caused unpromising market demand for creative products in Ipoh.

> While many are migrating outward, Ipoh now has limited working professionals, who usually have higher purchasing power and will more willingly pay for creative or added valued products. (Curator of a museum, interview, 21 September, 2019)

7.3 The Nexus Between State, Market and Civil Society

The Ipoh Local Plan 2020 and SAP of Ipoh Town-Tin Heritage City 2020, formulated in 2012 and 2014 respectively, indicated Ipoh City Council's adoption of conservation-centric regeneration strategy through creative place-making initiatives. The initiatives are to rebrand Ipoh city centre as a tin heritage city. Prior to the formal execution of these plans, an informal and private-led creative redevelopment project in the old town has somewhat triggered the 'creative pulse' in ITHC. It consequently nurtured creative urban entrepreneurism and motivated other private engagements in creative initiatives. Private stakeholders and investors started to invest in creative place-making initiatives towards harnessing commercial spin-off dividends through local tourism. These complemented and supported the local council's efforts in executing the planned creative strategies. The structured projects led by Ipoh City Council with support from the state government and the private sector (i.e. heritage property proprietors, developers, corporates) progressively built up the local creative art and cultural scene within ITHC. Seemingly, Ipoh city is following George Town's creative city-making footsteps but Ipoh is still at the stage of building up its creative contents and establishing its own creative ecosystem.

Since 2018, with a more specific vision to transform Ipoh into a UNESCO Creative City, the state government actively spearheaded creative initiatives. In 2018, the State EXCO for Tourism, Art and Culture declared his vision to turn Perak into an international destination through a strategic tourism master plan, pivoting local sustainability, identity, creativity and variety. During the period of his portfolio, he actively drove PORT into playing its role to nurture and showcase Perak's creative talents in diverse art fields. As part of his planned strategies, he was keen to promote Perak as a state of festivals.

> We attempt to reposition Ipoh. Working together with the local government and our creative agency (PORT), we are keen to develop Ipoh as a creative city with gastronomy as a way forward. We are also looking into the local art scene. We initiated some local programs and created platforms to link and gather our local artists, creative experts, active practitioners as well as the art spaces and the private stakeholders involved. We are planning to establish our Ipoh City gallery and allocate space for artists to showcase their artworks. (Perak State EXCO for Tourism, Arts and Culture, interview, 28 September 2019)

For the art sector, the State EXCO launched *Dana Kreatif Perak* (Perak Creative Funds) in November 2019, under the management of PORT. It aims to support Perak's creatives (aged 18 to 45 years old) in visual arts, performing arts and music towards developing the local creative industries, particularly in art. As a start, *Dana Kreatif Perak* is limited to RM100,000. Despite a limited funding amount if compared to funding available to KL and Penang's creatives, it is indeed an encouraging step towards driving the local art and creative sectors. In its first year, 25 local creatives out of 78 applicants received funding from *Dana Kreatif Perak* in January 2020 (Saleh, 2020).

Aligned with Ipoh City Council's development strategy for ITHC, the creative initiatives are well supported by both the government and private sectors. However, the change of the ruling government in Perak since April 2020 has raised a concern about whether the new government is still interested in the creative city agenda. Yet, according to the restructured State EXCO member's new portfolios, the previous portfolio of Tourism, Arts and Culture has been abolished while 'tourism' is now under the State Exco of Housing, Local Government, Public Transport and Tourism (www.perak.gov.my). Thus, art and culture will most likely not be the focused aspects of the current state government. Instead, the current State Exco is now wearing all four very heavy-weighted hats (i.e. 'Housing', 'Local Government', 'Public Transport' and 'Tourism'). Each carries different scopes of roles and functionalities under different ecosystems. Therefore, it poses a doubt about the rationality of the current state government's development approach. The shifted direction and focuses of the state government may consequently affect the existing adoption of creative city strategy. It may probably weaken the public-private partnership that has been built-up by the previous government. Consequently, it may challenge the sustainability of existing creative projects and initiatives in ITHC.

As conclusion, the creative pulse in Ipoh city centre was initially triggered by an informal private-led initiative prior to the formal execution of Ipoh City Council's planned creative strategies. Along the development process, the local council's creative strategies were supported by more private engagements and active state government intervention in the creative city agenda. Nevertheless, all government-initiated and private sector creative projects that have been accomplished thus far are just marking an initial phase of building Ipoh's creative scene. As way forward, the current state government needs to understand and reaffirm its commitment to the creative city agenda by establishing effective and long-term public-private-community partnerships as a means to build a sustainable and inclusive creative ecosystem for Ipoh.

References

Bunyan, J. (2018, June 30). *Private Museums All the Rage in Perak*. Malay Mail. Retrieved from: https://www.malaymail.com/news/malaysia/2018/06/30/private-museums-all-the-rage-in-perak/1647101. Accessed 3 Oct 2020.

Clark, D. (1989). *Urban Decline: The British Experience*. Routledge.

DoSM (Department of Statistics Malaysia). (2019, October). *Malaysia Statistical Handbook 2019*. Department of Statistics, Malaysia.

DoSM (Department of Statistics Malaysia). (2020a, August). *Poket Stats Negeri Perak ST2 2020*. Department of Statistics Malaysia.

DoSM (Department of Statistics Malaysia). (2020b, July). *Household Income & Basic Amenities Survey Report by State and Administrative District – Perak 2019*. Department of Statistics, Malaysia.

Encyclopaedia Britannica. (n.d.). *Ipoh, Malaysia*. Encyclopaedia Britannica online. Retrieved from: https://www.britannica.com/place/Ipoh. Accessed 3 Oct 2020.

ICC (Ipoh City Council). (2014). *Ipoh Heritage Tin City Special Area Plan 2020 – Final Draft (Rancangan Kawasan Khas Pekan Ipoh – Bandar Warisan Bijih Timah 2020 – Draf Akhir)*. Ipoh City Council (ICC).

ICC (Ipoh City Council). (2016, July 22). *Lonely Planet Names Ipoh as the Sixth Hottest Place to Visit in Asia*. Retrieved from: http://www.mbi.gov.my/en/mbi/media-centre/news/lonely-planet-names-ipoh-sixth-hottest-place-visit-asia. Accessed 3 Oct 2020.

ICC (Ipoh City Council), & JPBD (Town and Country Planning Department Perak Darul Ridzuan). (2012, July 5). *Ipoh Local Plan 2020 (Rancangan Tempatan Ipoh 2020) Jilid 1 Strategi dan Candangan Pembangunan (Volume 1 Development Strategies and Recommendations)*. Ipoh City Council (ICC) & Town and Country Planning Department Perak Darul Ridzuan.

Khoo, S. N., & Lubis, A. R. (2005). *Kinta Valley: Pioneering Malaysia's Modern Development*. Perak Academy, Ipoh.

Lee, A., Foo, H. P., & Choong, C. (2014, January 16). *Ipoh – When "Dressing Up in Tin" Will Still Be King?* Ipoh Echo. Retrieved from: https://www.ipohecho.com.my/v4/article/2014/01/16/ipoh-when-dressing-up-in-tin-will-still-be-king. Accessed 4 Oct 2020.

Loh, I. (2014, June 30). *Artist Splashes the Walls Around Ipoh Old Town with His Renowned Artwork*. The Star Online. Retrieved from: https://www.thestar.com.my/News/Community/2014/06/30/Coffeecommissioned-murals-Artist-splashes-the-walls-around-Ipoh-Old-Town-with-his-renowned-artwork/. Accessed 4 Oct 2020.

Saleh, E. (2020, January 26). *Pemberian dana Kerajaan Perak kepada industri kreatif wajar dicemburui – penggiat seni*. Perak Insights. Retrieved from:

https://perakinsights.com/2020/01/26/pemberian-dana-kerajaan-perak-kepada-industri-kreatif-wajar-dicemburui-penggiat-seni/. Accessed 10 Oct 2020.

Steinberg, F. (1996). Conservation and Rehabilitation of Urban Heritage in Developing Countries. *Habitat International, 20*(3), 463–475.

The Star. (2014, June 17). *Modernisation at a Cost.* The Star Malaysia. Retrieved from: https://www.pressreader.com/malaysia/the-star-malaysia-star2/20140617/281530814095732. Accessed 10 Oct 2020.

Tunyaporn, H. (2018, August). *Ipoh: An Attempt to Bring Back Life of Malay Peninsula's Once Thriving Economic City.* TunyaArticle.wordpress.com. Retrieved from: https://tunyaarticle.wordpress.com/2018/08/20/ipoh-an-attempt-to-bring-back-life-of-malay-peninsulas-once-thriving-economic-city/. Accessed 4 Oct 2020.

Wee, W. L. H., Yoon, D. K. T., & Guan, G. G. G. (2014, March). Revitalisation of the Old Township of Ipoh, Malaysia. *Journal of Place Management and Development.*

CHAPTER 8

Johor Bahru – A Border City's Ideals and Challenges

Abstract This chapter discusses the interesting role of Johor Bahru (JB) as a border city between Malaysia and Singapore as the city aspires to be a creative and cultural city in Malaysia's southern region. Johor Bahru role's will also be dissected against Johor state's aspirations of developing Iskandar Malaysia. The chapter will also provide some insights on the relationship between the state, private sector and civil society in adopting the creative city strategy to regenerate downtown Johor Bahru.

Keywords Creative city • Urban regeneration • Johor Bahru • Iskandar Malaysia • Border city

8.1 Background of Johor Bahru

The city of Johor Bahru (JB) was founded by Temenggong Daeng Ibrahim in 1855 by the name *Tanjong Puteri* after the sovereignty rights of Johor (excluding Kesang in Muar) was ceded to him by Sultan Ali Ibni Sultan Husain Syah on 10 March 1855. Hence, a new administration centre was developed in Tanjong Puteri by Temenggong Daeng Ibrahim who originated from *Telok Belanga*, Singapore. He established an office at *Bukit Bendera* (currently known as *Bukit Timbalan*), and later, the new administration was renamed as *Iskandar Puteri*. His son, Sultan Abu Bakar, later succeeded the throne upon Temenggong Ibrahim's demise on 31 January 1862. *Iskandar Puteri* flourished under Sultan Abu Bakar's reign with

S. L. Khoo, N. S. F. Chang, *Creative City as an Urban Development Strategy*, https://doi.org/10.1007/978-981-16-1291-6_8

growing population and booming socio-economic vibrancy. The date 1st January 1866 marked a defining moment when Sultan Abu Bakar officially opened the *Istana Besar* and *Iskandar Puteri* was ceremonially renamed as JB (IRDA & JBCC, 2013).

Sultan Abu Bakar, acknowledged as 'Father of Modern Johor', was instrumental in developing and modernising JB into a town with the construction of prominent buildings like the Royal Palace, the Sultan Abu Bakar Royal Mosque and the Chief Minister's residence. The construction of these buildings was mostly by Wong Ah Fook, a Chinese contractor who later fostered close ties with Sultan Abu Bakar. That period also witnessed an exodus of Chinese immigrants and the myriad of buildings built for the Chinese community back then are still present today (Yap, 2016).

Modern day JB is capital city of Johor, the state located at the southernmost region of Peninsular Malaysia. With a population of about 1.4 million in the city (2014) and close to 2 million in the metropolitan area (Iskandar Malaysia), JB is the second most populous city in Malaysia (Yap, 2016: 7). In 2019, Johor state registered an estimated population of 3,771,100 representing 11.5% of the national population (DoSM, 2020a). As per 2019 data, a household in Johor earns a median monthly income of RM6427 and mean monthly income of RM8013 (DoSM, 2020b). Overall, Johor registered 3.9% incidence of absolute poverty (DoSM, 2020b). Johor's economic performance relies on services sector, manufacturing and agriculture which contribute 49.3%, 29.4% and 12.4% respectively to Johor's GDP in 2018 (DoSM, 2020a).

As the state closest to Singapore, Johor is constantly reinventing itself to be competitive nationally and regionally. In 2006, the Iskandar Malaysia (IM) economic region was launched within Johor and IM was earmarked to be the preferred international metropolis to invest, work, live and play. About twice and thrice the size of Hong Kong and Singapore respectively, IM encompasses 221,634 hectares of land area within the southern-most part of Johor. The birth of IM, however, has resulted in two huge transformations. First, the drastic hike in urban land prices caused agglomeration of activities in certain areas; and second, the rise of new socio-economic groups (i.e. middle-class professionals, managers) with certain lifestyles and preferences (Embong & Macmillan, 2002) (cited in Razak et al., 2016: 124). While IM has attracted investment and witnessed spectacular physical developments under Iskandar Regional Development Authority (IRDA), the exodus of affluent residents and agglomeration of businesses in specific areas like the special economic corridor caused certain locations to be exclusive, elitist and has priced out low-income residents to places with fewer services and amenities (Richard, 2014) (cited in Razak et al., 2016: 124). Currently, the alarming concern pertains to IM as the billion-dollar failure where there are too much property oversupply and not enough people, particularly talent, to drive growth and sustain the place (Wah, 2020). Population growth has been sluggish with just 350,000 people in 12 years. This implied a small domestic labour force that is unable to cope with shortages (Pennington, 2017).

Thus, the study of JB in this book is timely and in line with national aspirations like Khazanah National Berhad's commitment to develop Iskandar Malaysia (IM) through IRDA. Subsequently, Khazanah's subsidiary Think City was co-opted (since 2016) to intervene as urban regeneration mechanism to revitalise downtown JB. In the Comprehensive Development Plan (CDP) of IM, there are five designated 'Flagship Zones', namely, JB City Centre (Flagship Zone A), *Iskandar Puteri* (Flagship Zone B), Western Gate Development (Flagship Zone C), Eastern Gate Development (Flagship Zone D) and Flagship Zone E that covers Senai and Skudai, Senai International Airport and MSC Cyberport City. Refer to Fig. 8.1 below to view the Flagship Zones of IM. This chapter earmarks Zone A specifically (i.e. JB downtown) in the quest to explore the city as a viable creative city.

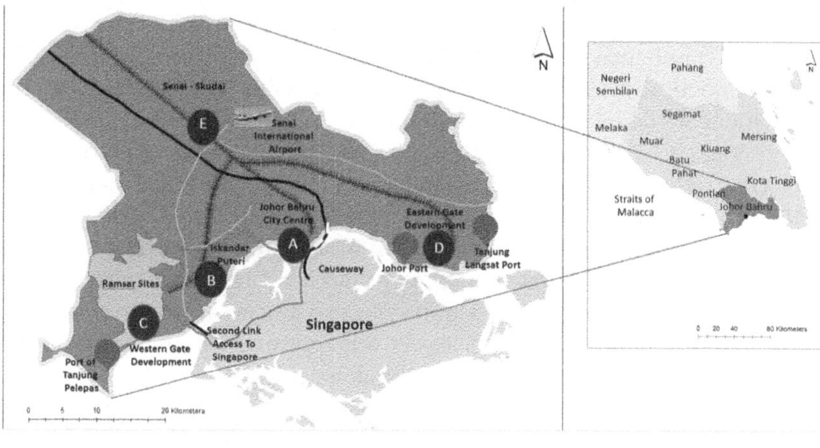

Fig. 8.1 Flagship Zones and location of JB city centre within Iskandar Malaysia. (Source: Authors' own map)

In the Iskandar Malaysia Comprehensive Development Plan 2005, it specifically positioned JB as a focal point for heritage, culture, urban tourism, education, creative services, retail, administrative and financial services. This positioning was further consolidated in the Johor Strategic Growth Plan produced by the Johor State Government. Concurrently, the JB City Council also has intentions, through their JB Transformation Plan, to position downtown JB as a tourism and retail centre besides reinvigorating the city's creative and cultural endowments (TCSB, 2019: 41). Extending from existing urban regeneration works by IRDA, Think City and JB City Council, the process of dissecting the promise and potentials of downtown JB (Flagship A) will then facilitate support and establish better linkages with Flagship B (*Iskandar Puteri*) (TCSB, 2019: 5) as a means to address the aforementioned 'stunted growth' in Iskandar Malaysia.

The plan by Think City proposes a two-phase strategy involving urban regeneration in downtown JB first and then followed by place-making activities in Medini, *Iskandar Puteri*. The rationale by Think City is simple and with the anticipation that regeneration ought to begin in brownfield sites like downtown JB where there is enough population density, ongoing economic activities and existing cultural heritage to catalyse population movement. It is projected that the momentum derived from activities in downtown JB would eventually expand and extend to greenfield developments in Medini/*Iskandar Puteri*, extending life and vibrancy to these new townships. Four key interventions in downtown JB include: (i) enhance liveability of downtown JB by collaborating with JB City Council and local stakeholders to regenerate JB City Centre; (ii) implement social inclusion programmes to enhance inclusivity at all societal levels; (iii) collaborate with public and private sector to strengthen JB's creative entrepreneurs' ecosystem; (iv) engage and build stakeholder relationships (TCSB, 2019: 13). To date, while there are signs of improvements in regenerating downtown JB but more can be done to unleash the city's creative and cultural potentials which are latently in existence but largely underexplored due to a non-cohesive and weak creative and cultural ecosystem in the city. This aspect was repeatedly highlighted during interviews with key informants. Looking specifically into downtown JB, the area is merely 119.8 ha with a relatively small population size of 3242 residents who are predominantly from low-income households. See Figs. 8.2, 8.3 and 8.4 for the map and pictures of downtown JB. According

Fig. 8.2 Plan of downtown JB. (Source: JB City Council, 2020)

Fig. 8.3 A scene of downtown JB. (© Suet Leng Khoo (August, 2019))

Fig. 8.4 Scene of JB historic core. (© Suet Leng Khoo (August, 2019))

to Think City's 2016 Baseline Study, it was reported that downtown JB presently serves as an administrative and retail centre, with minimal creative, cultural and knowledge-based activities. The downtown area also has substantial amounts of underutilised spaces and vacant floorspaces/areas of over 712,520 sq. ft. In comparison to downtown Kuala Lumpur and George Town World Heritage Site, downtown JB has more spaces between buildings with just 32% as building footprint compared to 33.5% and 48% respectively in Kuala Lumpur and George Town (TCSB, 2016: 1, 4, 59).

8.2 Situation Analysis of Creative City Strategy for JB

The subsequent sections will discuss JB's scenario in embracing the creative city strategy.

8.2.1 Unleashing JB's Regional Role in the Big Picture

The choice of examining JB as a potential Malaysian creative city invited many surprises and scepticisms. Concerns pertaining to *'What is there in Johor that is creative and worthy to be branded as creative?'* often cropped up as the opening response by interviewees until they gradually comprehend the gist of the concept as the interview progressed. Undeniably, every city has their creative side but subject to discovery or recognition; or, in other cases, dictated by the development trajectory of the city in terms of unleashing those creative elements and acknowledging them. In JB's case, while the city might pale in its shadow against George Town which is a UNESCO Heritage City and KL the capital city, JB possesses a 'border city' role adjoining Singapore. This position can potentially be leveraged to elevate the city to greater heights at the regional level and possibly further afield. As neatly surmised in Think City's JB Programme Phase 1 Report, the micro urban role of JB ought to be positioned within a broader big picture depiction with 'the recognition of JB as a pivotal city in the development of Malaysia's southern region, given its location at the international border of Malaysia and Singapore, proximity to Iskandar Malaysia and regional positioning within the Malacca Straits diagonal connecting Indonesia, Malaysia and Singapore' (TCSB, 2019: 13).

From a regional development perspective, JB's role is also viewed in terms of two (2) regional planning endeavours, namely, (i) the ASEAN Supercity; and (ii) the Malacca Straits Diagonal. The ASEAN Supercity regional development is an extension of the existing Indonesia-Malaysia-Singapore Growth Triangle (IMS-GT) concept formalised back in 1994. The IMS-GT was mooted to advocate cross-border collaboration and development amongst the three countries. It was also anticipated that by leveraging each territory's competitive advantages and by fostering interactions amongst these countries, it would also spur urban growth and development. The current ASEAN Supercity extended from IMS-GT to encourage more significant integration between the three regions, underpinned by access to railway, port and road to ASEAN's burgeoning market of over 600 million people. The second regional initiative is the Malacca Straits Diagonal which is a good option to steer Malaysia's competitiveness. Beginning from the southern border of Thailand and spanning the west coast of Peninsular Malaysia to Singapore, this region has three super cities (Greater KL, George Town conurbation and Iskandar Malaysia) that comprise more than 60% (16–18 million) of the nation's population. In cognizance of the interconnectedness of an urban system, these three cities are envisioned to generate and sustain a distinctly integrated urban system that will be more innovative and competitive as a collective whole (i.e. system) instead of disaggregated as sole cities (TCSB, 2019: 21).

8.2.2 Reinstating and Acknowledging JB's Creative and Cultural Endowments

Just like any other cities, JB has its own share of creative and cultural endowments to showcase though some are not accorded due recognition. These assets ought to be acknowledged as prerequisites in creative city-making. As part of the broader JB Transformation Programme mooted in 2010, strong state intervention in the form of strategic partnership between the JB City Council and Think City (which made its presence in 2016) kickstarted an initiative to map out the heritage and cultural assets in JB's Coronation Avenue. Located along the avenue are heritage jewels and iconic monuments like the Sultan Abu Bakar Mosque, Johor Ancient Temple, Arulmigu Rajamariamman Devasthanam Temple (see Fig. 8.5), Gurdwara Sahib Sikh Temple, Catholic Church of Immaculate Conception and *Masjid India JB*, signifying religious, ethnic and racial eclecticism of the city. These institutions and monuments are indeed heritage gems that form part of the physical infrastructure of JB as a creative and cultural city.

JB is also gradually gentrifying with the coming of new cultures especially in retail, gastronomy and the presence of street art/murals. While JB still has century-old establishments like Salahuddin Bakery and Hiap Joo Bakery (see Figs. 8.6 and 8.7), there are signs of hipster coffee bistro culture entering the city as shown in Figs. 8.8 and 8.9.

A focus group discussion with city council officials exposed an interesting fact. Apparently, development plans like the Conservation Special Area Plan for JB never specifically assigned a dedicated section or agenda as efforts towards creative city-making, but existing initiatives like heritage trails/walks (see Fig. 8.10) and selected urban events are in fact creative and cultural endeavours. According to the official, these initiatives have been there and functioned as merely *'fillers'* to fill the gaps in the report as ways to invigorate the city. Arguably, efforts are underway but urban managers and stakeholders are unaware of the creative city concept thus their inability to link their existing efforts to the creative city agenda and conceptualise the bigger picture.

Fig. 8.5 The Arulmigu Rajamariamman Devasthanam Temple. (© Suet Leng Khoo (August, 2019))

Fig. 8.6 Salahuddin Bakery still uses wood fire for baking. (© Suet Leng Khoo (August, 2019))

> [The creative initiatives] are not direct but act as fillers in programmes. If we look, we see it but it is not presented in the form of write-up. People are unclear about these creative initiatives. These initiatives are there but no direct impact, but it will attract tourism and generate returns to the economy. (JB City Council official, FGD, 6 August 2019)

However, development plans like the Ibrahim International Business District Plan (JBCC & JCorp, 2016) and JB Transformation Plan do contain key phrases to transform JB into a 'heritage and cultural city'. Similarly, the 2025 Tourism Special Area Plan for Johor Bahru City also strategised to position JB city as an 'Urban Heritage & Lifestyle' locale. These phrases in the development blueprints are inkling of JB's strategic direction towards becoming a culture-led creative city. Arguably, just like most Asian cities, the cultural turn in JB is somewhat driven by the state.

Fig. 8.7 Traditional oven in Hiap Joo Bakery (since 1919). (© Suet Leng Khoo (August, 2019))

Additionally, the state of Johor seemingly differentiates itself from other Malaysian states by upholding and safeguarding *'Bangsa Johor'* – the unique character, identity and culture of all de facto *Johorians*. During the interviews, the *'Bangsa Johor'* identity was highlighted occasionally by key informants who are themselves proud *Johorians*. JB, arguably, has its hidden and latent culture and heritage that can be leveraged but it has not been harnessed optimally. And the lack of creative energy and pulse within JB made the city assume a supporting role and play second fiddle to Singapore as highlighted here:

> There are initiatives but if you're unaware, you won't see it. Here, the NGOs are not as vocal as those in Penang. We have the galleries in Johor, they can't survive for Johor but they support the Singapore market. (JB City Council official, FGD, 6 August 2019)

Fig. 8.8 Public art-cum-mural on exterior wall of café in refurbished shophouse. (© Suet Leng Khoo (August, 2019))

Fig. 8.9 Modern art gallery cafes like Senso are mushrooming in downtown JB. (© Suet Leng Khoo (August, 2019))

8.2.3 Non-cohesive Ecosystem and Lack of Supporting Mechanisms

While JB has hidden creative and cultural gems, the vibrancy of cultural activities is lukewarm and the city lacks a supporting system (i.e. human capital, funding) to activate and drive the creative industries even though Khazanah (through Think City) has intervened and indications of progress is underway. This plight was repeated multiple times by the authorities and a local think-tank.

> Tourists don't see our creative endeavours when they come to JB. Penang is lucky because of NGOs and tourist guides who mobilize the industries. We do not have supporting activities. Malacca and George Town have UNESCO. In JB, who will guide us? We guide ourselves. We can't implement like Malacca because money (funding) does not come to us like Malacca. We also don't have special financial allocations. Everything is pri-

Fig. 8.10 The Tan Hiok Nee Heritage Walk Arch in downtown JB. (© Suet Leng Khoo (August, 2019))

> vate initiative. In the Local Plan, there's no allocation (funding) for creative activities. (JB City Council official, FGD, 6 August 2019)

> JB Transformation Programme always had this tagline of 'JB as Heritage & Cultural City' but when we did the baseline study, we found out that the heritage and cultural activities were less than 10% … or something. That means that they are not enough mechanisms and supporting elements in the city to say that the city is heritage and cultural city. Sadly, because of lack of mechanism … the creative community shy away from the city. You couldn't bring them back. The artist community feels very restrained. Restrictions on their movement. They don't have any spaces. They operate by themselves. They go to KL. (Urban regeneration think-tank, FGD, 6 August 19)

This is compounded by the fact that the multi-tiered government administrative structure has complicated matters when local authority action plans sometimes do not align with those at the Johor state level.

8.2.4 A Border City's Accidental Homelessness

As Landry (2008) highlighted, a creative city would require both hard infrastructure (i.e. buildings, roads, amenities) and soft infrastructure (i.e. people and their cultures, values). Despite having both infrastructural types though not developed and leveraged optimally, old historic cores like JB are still plagued with urban issues like urban poverty, traffic congestion and haphazard urban planning as disclosed in their Conservation Special Area Plan. To exacerbate the matter, an urban issue unique to JB was unearthed during the interviews. The issue of 'accidental homelessness' was highlighted and it has to do with JB's border city status and continued dependency on Singapore for employment. Due to higher currency exchange rate in Singapore, *Johorians* have opted to eke a living in Singapore and most of them are in the working class and they fill positions in Singapore's services industries (i.e. hotels, retails, F & B). These *Johorians* who cross the causeway to work daily will use JB downtown as their transit areas to shower, have their meals, rest or linger around. In the public's eyes, this group of people are akin to temporary 'homeless' people and they pose as an urban eyesore to JB though it is an unintentional phenomenon. This urban blight was shared by the interviewees as cited here:

> The issue here is maybe about 50,000 *Johorians* cross over the borders to Singapore every day to work. The jam takes about 2–3 hours. These people based on their qualifications still earn about RM4000–5000 but they are not from JB. They are from other Johor districts. They buy motorbikes. They work from early until late. Instead of going back home they use the public toilets and stay in City Square area (in downtown JB) until the city council started to build halfway houses. These houses are actually for people who don't have houses, cannot afford and those living on the streets. It is for the homeless but they have been used by these people. (Freelance cultural worker, interview, 5 August 2019)

> Our course we have homeless people. It's easy to tell. If you walk around and feel scared and unsafe, then that's evidence of homeless people lingering in the city. Old cities will have this issue. Penang, KL all surely have. In JB, we have homeless issue too. Johor is special…there's a lot of social activities. Poor homeless people are not many but there are many homeless people who are working in Singapore. It is a bit strange. They use JB as a transit point. They are Malaysian citizens so we can't arrest them. (JB City Council official, FGD, 6 August 2019)

8.2.5 *Brain Drain*

A key informant with an urban regeneration think-tank highlighted that JB has 'talent' and there are existing creative endeavours that are 'organic, grassroot driven and privately-funded' but many of them are hidden, latent and unknown. When these creative talents eventually make it big, they gravitate to bigger cities like Kuala Lumpur and cause a void and brain drain in JB as quoted below:

> We have 40 applications but most of them are not living here. They live in KL but they are *'orang Johor'* (*Johorians*). These are the challenges. (Urban regeneration think-tank, FGD, 6 August 19)

> One of the best artists came from JB but these people have migrated to KL because cannot carve a livelihood here. You can find the big names are local boys. We're just wondering why this is happening. You don't see art exhibited here. (Freelance cultural worker, interview, 5 August 2019)

Arguably, JB's border city status might have caused the city to pale in comparison to Singapore which is coincidentally a UNESCO Creative City of Design. It is disheartening to know that JB, despite having the niche prowess for furniture design and manufacturing, is losing out its competitive edge and creative talents to Singapore. The interviewees shared this plight:

> If you talk about creative economy, we have a lot of furniture designers, interior designers, architects … so these elements of creativity. This is the part that you need to empower the people to be here in JB and not be in Singapore. I have a lot of my friends … we organised stuffs. When I organised TedEx Sungai Segget (in JB), most of my friends are designers. Chinese communities are mostly designers. Even though they won awards, the exhibition was in Singapore and not here. The innovation of furniture … we don't have that one-stop centre for furniture here. Yes, like a brain drain. This is not that we don't have. We have a lot…From history, cultural, heritage, product design. (Urban regeneration think-tank, FGD, 6 August 19)

8.3 The Nexus Between State, Market and Civil Society

No doubt, just like Kuala Lumpur, George Town and Ipoh, the city of JB is keen to hop onto the bandwagon of the creative city agenda. While JB's potentials as a creative city were duly highlighted in Think City's technical reports and features informally (i.e. JB Conservation Special Area Plan) and formally (i.e. JB Transformation Plan) in the city's development blueprints, the pathway leading towards that end leans towards a hybridised strategy. The strategy combines both a Western model of 'culture-led urban regeneration' aimed at reinvigorating the potentials of JB's decaying historic core and the Eastern model that has a strong hand of state intervention. JB's culture-led urban regeneration strategies and initiatives are visible and encapsulated in the myriads of development plans (i.e. JB Transformation Programme, JB Special Area Plan) aimed towards enhancing JB's competitiveness, entrepreneurship and liveability in order to rise quick enough to execute its role as Flagship Zone A – the first out of five portions of Iskandar Malaysia and staged catalyst to have spillover effects and stimulate growth in Medini, *Iskandar Puteri* (Flagship Zone B). It is clearly stated in the blueprints that the 'heritage and culture of JB' (located in Flagship Zone A) and its linkage with the past is pivotal to develop the new township of *Iskandar Puteri* in Iskandar Malaysia's Flagship Zone B – which will take time to develop in terms of pulse and character hence its dependency on older cities in the vicinity to shape it. Sadly, this utopian and positive relationship between JB (Flagship A) and *Iskandar Puteri* (Flagship B) is not so straightforward and not unfolding as desired.

This is where the Eastern model of creative city development with strong state intervention has set in for both JB and *Iskandar Puteri* in a rather aggressive manner. The JB City Council and IRDA mooted the JB Transformation Programme in 2010 when the inner city was decaying and subsequently the role of Think City (Khazanah's urban regeneration and city-making agency) entered the scene in 2016. Many forms of interventions and strategies have been put into place to revitalise downtown JB, and there is visual evidence of physical changes and improvements. Revitalising the city through arts and culture is gaining momentum. For instance, an interview with a key official highlighted the role of arts to revitalise JB. His viewpoints are stated here:

> In general, Arts is one of the elements that has been identified to arrest the potential decline in urban areas. To be functional, it has to be integrated into local economic activities and provide jobs to the local communities as well as provide new opportunities and interactions. The urban structure of Johor Bahru has been classified as City Centre and it is expected to enable Urban Tourism, Arts and Culture, and Administration. Hence, the project dubbed Iskandar Malaysia Heritage, Cultural and Arts Trails has been rolled out. To showcase the untold story of Iskandar Malaysia, its history, heritage and culture, several historical places of interest as well as arts and culture shall be rejuvenated, recreated and developed. The key initiatives such as Johor Performance House, Cultural and Arts Centre Marketplace, Historical Trail, and Tourist Waterfront could promote urban sustainability, improve quality of life and advocate social inclusiveness for Iskandar Malaysia. Nevertheless, one should not deny the rate of growth for urbanisation would probably outpace the rate of creative proliferation. Hence, Iskandar Malaysia encourages the interdependence and shared resources among urban areas which would lead to improved efficiency and productivity. (Vice-President of regional development authority, interview, 6 August 2019)

To implement all these would require collective efforts from all stakeholders especially the local communities. Despite IRDA and Think City's attempts to introduce a more holistic, participative and bottom-up approach by engaging local communities and private initiatives in JB, the process is slow. Unlike George Town, there are minimal grassroots movements (i.e. NGOs and local communities) in JB and they are less vocal and tend to shy away from partaking in urban governance until Think City came along with baby steps to engage and build trusts with the local communities through projects like *Paso Barokah@Pasar Karat*, JB Film Grants, Flat Taman Jaya Regeneration and such.

In terms of private sector commitment, it is inaccurate to say that JB is devoid of such initiatives because as mentioned by key informants that there are indeed 'organic, home-grown, self-funded' creative endeavours. A good case in point is the Johor Arts Festival organised by the Johor Society for the Performing Arts (JSPA). It all begun in 1998 when a group of creative minds congregated and wanted to 'stir up some excitement in JB', and later in 2000 registered as a non-commercial organisation with a mission to promote, nurture, support, advance, create awareness and inculcate appreciation for the arts in Johor. The founders – a husband and wife team – who were unlearned and untrained in the arts initially faced multiple adversities on their credibility and the right to do arts but eventually broke all boundaries when they successfully organised the JB Arts Festival for the past decade by leveraging local and foreign creative talents alike.[1] However, the clientele might not be as big as those in George Town and Kuala Lumpur coupled with the fact that JB creatives tend to gravitate to bigger cities like Kuala Lumpur and Singapore due to JB's less developed creative ecosystem. This is a pressing issue highlighted repeatedly during the interviews. To address this, Think City had objectivised to consolidate the creative entrepreneurs' ecosystem in collaboration with the public and private sector (TCSB, 2019: 46). Unequivocally, the role of Think City as catalyst and mentor in JB's city-making and urban regeneration is instrumental, at least in the last five years. Since 2016, Think City has implemented 112 projects in JB of which 99 are live. Apart from being grantor, Think City's approach that includes management and partnership initiatives with stakeholders and the local communities are commendable and bearing fruits. For instance, Think City was invited to support *Yayasan Warisan Johor* (Johor Heritage Foundation) on the refurbishment of *Galeri Tenun* (Weaving Gallery); and JB City Council's invitation for Think City to work with them to transform the Meldrum Area, a famous location in downtown JB for street food. Think City was also asked to provide input to the working committee on producing a Creative Economy policy for Johor state (TCSB, 2019: 93). Such a bottom-up and inclusive model is proving to be friendly, acceptable and hopefully sustainable in the long run.

References

DoSM (Department of Statistics Malaysia). (2020a, August). *Poket Stats Negeri Johor ST2 2020*. Department of Statistics, Malaysia.

DoSM (Department of Statistics Malaysia). (2020b, July). *Household Income & Basic Amenities Survey Report by State and Administrative District – Johor 2019*. Department of Statistics, Malaysia.

IRDA (Iskandar Regional Development Authority), & JBCC (Johor Bahru City Council). (2013). *Draf Rancangan Kawasan Khas (Draft Special Area Plan). Konservasi, Garis Panduan dan Glosari Kawasan Warisan Pusat Bandaraya JB. Jilid 1: Laporan Pelan Induk dan Garis Panduan. (Draft Conservation Special Area Plan. Conservation Guidelines and Glossary of Heritage Area of JB Historic Core. Volume 1: Masterplan Report and Guidelines).*

[1] Peggy Loh's blog, Fifteen years of Johor Arts Fests. Source: http://ppunlimited.blogspot.com/2018/10/fifteen-years-of-johor-arts-fests.html (accessed on 18 June 2020).

JBCC (Johor Bahru City Council), & JCorp (Johor Corporation). (2016). *Ibrahim International Business District Blueprint.*

Landry, C. (2008). *The Creative City: A Toolkit for Urban Innovators* (2nd ed.). Comedia Earthscan.

Pennington, J. (2017, July 20). *Iskandar Malaysia Fails to Shine Even with Significant Investments from Singapore.* ASEAN Today. Retrieved from: https://www.aseantoday.com/2017/07/iskandar-malaysia-fails-to-shine-even-with-significant-investments-from-singapore/. Accessed 15 June 2020.

Razak, M. R., Johar, F., & Khalil, R. A. A. (2016). The Impact of Iskandar Malaysia Development on Urban Amenities in Planning Malaysia. *Journal of the Malaysian Institute of Planners*, (Special Issue IV), 121–134.

TCSB (Think City Sdn. Bhd.). (2016). *Downtown JB Baseline Study.* Think City Sdn. Bhd.

TCSB (Think City Sdn. Bhd.). (2019). *JB Programme. Phase 1 Report: February 2016 – February 2019.* Think City Sdn. Bhd.

Wah, V. (2020, January 19). *Iskandar Malaysia: On the Edge of Glory or a Billion-Dollar Failure?* ASEAN Today. Retrieved from: https://www.aseantoday.com/2017/01/iskandar-malaysia-on-the-edge-of-glory-or-a-billion-dollar-failure/. Accessed 15 June 2020.

Yap, Z. H. (2016). *Johor Bahru: My City, My Heritage: A Sketch Book of Historical Buildings in Johor Bahru.* Galeri Hanzhen.

CHAPTER 9

Concluding Thoughts and Way Forward

Abstract This chapter will highlight key findings from the four case studies of Kuala Lumpur, George Town, Ipoh and Johor Bahru in terms of their uniqueness, readiness and viability in embracing the creative city strategy. The interrelationships between key development actors at the national and city level in adopting creative city development as a new form of urbanity will also be recapitulated. The chapter ends by offering some insights to craft a formulaic creative city development strategy that is all-encompassing, inclusive and sustainable for all Malaysian cities.

Keywords Creative city • Malaysian cities • Urban development • Inclusive • Sustainable

9.1 Looking at the Big Picture

As an aspiring developing nation, Malaysia's development trajectory is inevitably shaped by the globalisation process. As Malaysia strives to be economically competitive, the nation's economic base structurally shifted from industrial to services and now re-orientates towards the creative economy. Simultaneously, this shift has its bearing on Malaysian urban development when urban settings started to restructure in tandem by leveraging creative and cultural industries to drive urban development and regeneration. This book has showcased the scenarios in four selected Malaysian cities (i.e. Kuala Lumpur, George Town, Ipoh and Johor Bahru)

S. L. Khoo, N. S. F. Chang, *Creative City as an Urban Development Strategy*, https://doi.org/10.1007/978-981-16-1291-6_9

where each city attempted to dissect the compatibility of cultural and creative industries/endeavours with their respective urban (re)development policies.

The first case study illustrated the scenario in capital city Kuala Lumpur. The formidable way urban neoliberalism is unfolding has inadvertently earmarked KL as Malaysia's de facto creative city. In the process, KL is also striving to portray itself as a world class city with a global outlook. In due course, massive physical developments happening in Greater KL and the suburbs have hollowed out downtown KL causing certain parts of the historic core to become haunts for homeless people and migrant workers. To revitalise downtown KL, strong intervention from state bodies (i.e. Kuala Lumpur City Hall, MOTAC) and state urban regeneration body (i.e. Think City) are underway to transform downtown KL into a creative and cultural district. Concurrently, there are emerging private sector initiatives to revitalise KL where old shophouses, *godowns* and flats are refurbished and adaptively reused, as the case with *Godown* and Zhongshan Building, when second-generation property owners start to show interest and move back to the city. The gentrification process unfolding in KL is somewhat welcomed in this sense though the underlying capitalistic notions of gentrification need to also be understood and dealt with cautiously to deter social exclusion and urban inequality from worsening in KL. The presence of urban regeneration body, Think City, has also fostered fruitful private-public partnership evidenced through the case of Zhongshan Building. While KL's authorities and key stakeholders recognise that creative and cultural industries can be fused into the city's urban policy and re-development agenda, more concerted efforts have to go towards fathoming the real benefits and disbenefits of the concept given that 'creative city' has diverse meanings to city authorities, private sector and civil society.

In George Town, although there are no official labels to formally brand the city as a creative city despite possessing the prowess and criteria (see Khoo et al., 2014, 2015), arguably, George Town's creative/cultural endowments and endeavours thus far point the city towards that direction. George Town's 2008 UNESCO inscription is a defining moment that altered the urban form, fabric and cityscape of the heritage site. The once hollowed out downtown George Town heralded a distinctive cultural turn that motivated George Town to partake culture-led urban regeneration as a means to showcase the city's rich cultural heritage. In due course, a vibrant creative scene has emerged in George Town backed by strong state intervention through funding and support for local art and cultural activities. Though George Town inclines towards a cultural-centric creative city orientation where the city's unique cultural heritage is leveraged to stimulate local economic development, the city is not spared from the vagaries of neoliberalist urbanism infiltrating the heritage site such as gentrification, over-tourism and concerns over commodification of culture and heritage for tourism revenue. The absence of a socially inclusive creative city masterplan/policy, the current non-cohesive creative ecosystem and the state's tourism-oriented creative agenda singularly or collectively pose a challenge towards the development of an inclusive and sustainable creative city model in George Town.

In Ipoh, the cultural turn emerged when the Ipoh local government committed to regenerate Ipoh city centre through conservation-centric place-making strategies. These are clearly outlined in the city's Special Area Plan (SAP). Ultimately, the SAP aims to brand Ipoh city as a 'Tin Heritage City' in line with the city's historic functions and economic activities. Some local private-driven efforts have complemented the local government initiatives towards setting up a creative art and cultural scene in Ipoh Tin Heritage City. With aspirations to develop Ipoh to become a globally recognised UNESCO Creative City, the state government has actively invested in art and cultural projects since 2019. This aspiration marked an initial phase of establishing the city's creative ecosystem. However, this move may be interrupted due to shifting strategic directions and focus of the newly appointed Perak's state government that might not advance the creative city agenda. Nonetheless, the viability of creative city strategy for Ipoh city very much depends on local appreciation of local creativity and culture as well as a more cohesive ecosystem to ensure urban sustainability in the long run.

As for Johor Bahru, the motivation to embark along the cultural and creative city route is motivated by broader strategic directions within Johor state, namely, the development of Iskandar Malaysia. To this end, the move to regenerate and transform downtown Johor Bahru into a 'Heritage and Cultural City' serves two objectives. First, in the Comprehensive Development Plan of Iskandar Malaysia, Johor Bahru City Centre (Flagship Zone A) has been earmarked to be revitalised to support and have spillover effects to Medini, *Iskandar Puteri* (Flagship Zone B). Second, the JB Transformation Plan also has plans to reinvigorate decaying JB by leveraging the city's creative and cultural endowments thus unleashing JB's potential as a creative and cultural city. Compared to

KL, GT or even Ipoh, the gentrification process in downtown JB is relatively slower where signs of adaptively reused boutique hotels, café bistro culture and murals/public artworks are slowly emerging but still at its infancy. While JB has a myriad of forward-looking development blueprints (i.e. JB Transformation Plan, Ibrahim International Business District Plan, etc.) to steer the city's direction, the hardware and software aspects are still lacking. Unlike more matured cities like GT and KL, the supporting system for creative endeavours in JB is low where there is a dearth in human capital, funding and proper infrastructure to elevate creative and cultural activities to a higher level. Nonetheless, in the last five years, more concerted efforts by the state through IRDA, Think City and JB City Council have managed to reach out and include various stakeholders, especially local communities, towards transforming JB into a 'Heritage and Cultural City' in its own right.

The case studies have shown the motivation and forces driving creative city development in the four cities. The adoption of the policies is guided by each city's aspirations at the national, regional and global levels as well as their hierarchy, importance and functionality in the urban order. As Malaysia's capital city, Kuala Lumpur views the adoption as a way to revitalise downtown KL and also exude a world class/global city outlook while George Town's cultural turn is somewhat incidental due to the city's UNESCO status. While to some extent Ipoh might be emulating some of George Town's cultural and creative pursuits, the city actually has clear aspirations to be a UNESCO Creative City – something that is not on George Town's agenda as yet. As for JB, the city is an emerging cultural and creative city due to Iskandar Malaysia's direction to leverage JB to support the Medini, Iskandar Puteri (Flagship Zone B) as well as plans to revitalise a decaying JB downtown area.

Though many Western scholars argued that creative city development that hinges on culture-led urban development (Landry, Bianchini) is a strategy to reinvigorate socio-economic conditions in post-industrial cities for local development, this logic however is less relevant to the Malaysia context. Overall, as a developing nation, most Malaysian cities have not reached the post-industrial or deindustrialisation situation akin to European or American cities. For instance, KL's interest in creative city policy is triggered by the decline of its historic core and its capital city role to portray a global outlook whereas George Town is motivated by its UNESCO branding to showcase the city's cultural and creative content. Notably, a double branding might further stimulate local economic development in George Town. Although the econo-centric orientation and urban entrepreneurialism motivations are playing out in all four cities, their executions are less dictated by market-centred policies. Instead they are driven by political imperatives with strong state intervention in all four cities albeit in varying degrees, forms and magnitude.

Additionally, the discourse of creative city cannot be detached from deliberating the intrinsic link between human capital and their motivation to stay in a particular urban locality. From the four case studies, arguably, creative city development in KL, GT, JB and Ipoh is fraught with issues of brain drain although manifested in different levels and magnitude. At the local/national level, secondary cities like JB and Ipoh find it difficult to retain and nurture their creative talents who leave for greener pastures in Kuala Lumpur. But, at the international level, Kuala Lumpur as Malaysia's premier city is also encountering brain drain of creative talents with more severe ramifications when Malaysian creatives find it challenging to carve a livelihood in KL and opt to move abroad. Once they leave Malaysian shores, it is difficult to entice them back thus creating a void of creative talents in Malaysia. As espoused by Landry (2008), a creative city can only be developed and sustained through a good mix of hard infrastructure and soft infrastructure in this case human capital. Until and unless a more conducive and enabling environment is fostered in all four cities to nurture creative talents, otherwise, the issue of brain drain will perpetuate or even be compounded.

9.2 The Way Forward

During the writing of this book (i.e. year 2020), Malaysia witnessed two major incidences that warrant a brief mention here. These incidences in turn provoked more questions than answers given that both incidences are currently unfolding and the final outcomes are far-reaching and uncertain. The precarity of both incidences have definitely impacted and implicated the way forward pertaining to creative city-making in Malaysia. First, Malaysia underwent a sudden change of government in March 2020 and alongside this a change in ministerial positions and possibly shift in visions, policies and strategies. The creative and cultural industries are impacted each time a new minister assumes office. Contingent upon the incumbent's acceptance, mindset and appreciation towards arts, culture and creativity, arguably, Malaysia's creative and cultural industries have a lot to reckon with during this transitional period. To aggravate the uncertainties that derived from this shift, just like any other countries, Malaysia also succumbed to the Covid-19 pandemic where the Malaysian government responded by imposing the Movement Control Order (MCO) from 18 March 2020 onwards. The current fluctuating cases reported have caused the government to impose Conditional Movement Control Order (CMCO) in selected Malaysian cities that are flagged as red zones indicating high Covid-19 cases. Under such circumstances, cities in Malaysia have been partially locked down. KL as the biggest Malaysian city where the hype of the creative and cultural pulse congregates is currently bearing the brunt of Covid-19. As highlighted by global reports (OECD, 2020; Hall, 2020), the creative and cultural industries globally are amongst the worst hit sectors due to Covid-19. In Malaysia, it is no exception where events, shows, concerts and performances have all been cancelled and brick-and-mortar cultural venues have been closed indefinitely due to Covid-19 and the need to socially distance from one another. For an industry that emphasises a lot of face-to-face interactions and physical display of their expression as medium to appreciate a particular creative work or cultural experience, undeniably, Malaysian creative and cultural workers are badly hit when they are laid off temporarily and indefinitely. Many have been asked to 'Work from Home' as an alternative where they leverage on internet and communication technology (ICT) to stay afloat.

Addressing this vagary, CENDANA was quick to respond by conducting a baseline study to have a snapshot understanding about the situation and fate of Malaysia's creative and cultural industries (CCI). With arts, culture and creativity often being side-lined in the nation's broader development agenda, the fate of CCI and its employees' ability to survive and brace through this unprecedented and challenging time has in fact provoked another question regarding the fragility and resilience of a creative city and creative economy to withstand exogeneous and endogenous shocks. This, undeniably, would be a challenging yet opportune time for aspiring Malaysian creative cities to gauge their preparedness and address their weaknesses. To begin, the four case study cities in this book should endeavour to build resilience and live up to its definition as purported by Landry (2008) where creative cities can organically and eventually derive at solutions to face any adversities, in this case, the on-going pandemic and its aftermath. Not only does a creative city need to be imbued with elements of creativity, culture and be solution-based, more importantly, unexpected shocks like Covid-19 highlight the fact that creative cities also need to be inclusive, collaborative and garner the aggregate intelligence and collective creativity of all and sundry in the urban milieu to combat Covid-19 and whatever future shocks that might inflict a city. To this end, Malaysia's pluralism and multiplicity of cultures, ethnicities and creativity embedded within the urban milieu will augur well and give Malaysian creative cities a better head start provided concerted efforts are partaken among key stakeholders and urban managers.

The writing of this book has triggered reflections in four (4) areas related to creative city development in Malaysia, namely, (i) concept and taxonomy; (ii) urban planning and development; (iii) regional planning and development; (iv) development studies. From the reviews and critiques in Chaps. 2, 3 and 4, arguably, the concept, classification and taxonomy of creative city cannot be directly uprooted from any Western or Eastern country to fit Malaysian cities. Careful understanding of the upsides and downsides of the concept and taxonomy, its underlying agenda/motivation and its applicability to Malaysian cities are key considerations that Malaysian key stakeholders should take note in order for the strategy to generate desirable outcomes. The concept should be localised and be tailored to the Malaysian context by taking into account local idiosyncrasies and nuances of each Malaysian city. There should not be a direct import or a 'one-size-fits-all' generic concept to overarch the rhetoric and plans towards creative city-making in Malaysia.

Since time immemorial, urban planners and architects have been responding to the emergence of new diseases (i.e. Spanish flu, plague, etc.) and correspondingly the pressures and demands of public health to replan and retrofit the built environment. The current Covid-19 pandemic is no different where urban planners and designers will have to be exceptionally creative in reconfiguring, repurposing and reimagining a future creative city as the Covid-19 pandemic permeates into the urban realm. Landry's definition of hardware and software in a conventional creative city would be radically different following the aftermath of Covid-19. The future of brick-and-mortar cultural infrastructures (i.e. theatres , galleries, museums) and in-person cultural activities will need to be thoroughly reconsidered as the creative industries and creative employees embark new ways of producing, consuming and disseminating culture and creativity, predominantly facilitated and enabled by ICT (i.e. social media, augmented reality, virtual reality platforms, etc.). In Malaysia, these considerations should be incorporated into development blueprints by state officials and creative city vanguards as they endeavour towards constructing resilient and proactive creative cities that can brace future pandemics and calamities.

From the literature in Chap. 3 and also evidenced through this study, the spillover effects of creative city development transcend a city's boundaries and can advance to a regional/national level, thus, creative mega urban regions. No doubt, this book has shown that the four selected case cities of KL, GT, Ipoh and JB could collaborate (instead of compete) and cross-fertilise their respective creative and cultural contents at the national level as well as elevate the strategic partnership to a higher regional level among ASEAN creative cities found in Singapore, Indonesia and Thailand. For instance, George Town (with the city's gastronomic prowess) could consider collaborating with Thai city of Phuket which is a Creative City for Gastronomy. Likewise, Johor with homegrown talent in design should collaborate with Singapore which is a Creative City of Design. National collaboration across Malaysian cities and regional partnerships/joint-ventures among ASEAN cities should be the 2-prong strategy and way forward. This will augur well for all aspiring Malaysian creative cities as they unleash their own potentials as well as enhance their viability and visibility on the national/global platform.

Finally, some key food for thoughts for Malaysian creative cities depart from the key tenets found in the field of development studies. Evidences from the four case studies have showed instances on urban neoliberalism and consequently gentrification as well as incidences of social exclusion, homelessness and urban poverty. Arguably, these urban issues are the downsides of the creative city concept as highlighted by scholars (Pratt, 2011; McCann, 2007). Acknowledging this, Japanese scholar Sasaki (2010) has reconceptualised and coined a 'socially-inclusive creative concept' that addresses the aforementioned urban issues (i.e. poverty, homelessness, etc.) in Japan. At this juncture, Malaysian planners should take heed of such issues and address them accordingly. Urban managers should collaborate with scholars to reconceptualise a Malaysian model of creative city that propagates growth with equity by advocating a bottom-up and participative approach where all development actors (i.e. state, private sector, civil society, etc.) are included and involved in constructing their own model of a resilient, inclusive, proactive and sustainable creative city. In line with SDG No. 11 and the New Urban Agenda's tagline of leaving no one behind, the clarion call to embed urban inclusion into all development blueprints should underpin the foundations of a socially inclusive creative city in Malaysia. While acknowledging that these global mantras are overarchingly huge in nature, the process to adopt, adapt and translate them from the global to the local level will involve inclusive urban governance, effective public administration and integrated urban planning where every urban citizen's needs, rights and voices in Malaysia have equitable opportunity to be heard and acknowledged. In such times of turbulence and change in Malaysia, the creative city strategy will have to be malleable because the meanings attached to the narrative and discourse will be constantly revisited and reconstructed to respond to the particular policy agenda, local needs and aspirations of each Malaysian city which is distinctively unique to the local urban setting.

References

Hall, S. (2020, October 21). *COVID-19 Shows It's Time to Rewrite the Script on Jobs in the Creative Economy*. World Economic Forum. Retrieved from: https://www.weforum.org/agenda/2020/10/covid-19-shows-it-s-time-to-rewrite-the-script-on-jobs-in-the-creative-economy/. Accessed 19 Nov 2020.

Khoo, S. L., Badarulzaman, N., Samat, N., & Dawood, S. R. S. (2014). Capitalising on Urban Cultural Resources for Creative City Development: A Conceptual Review and the Way Forward for Malaysia's George Town. *GEOGRAFIA Online Malaysian Journal of Society and Space, 10*(5), 20–29.

Khoo, S. L., Samat, N., Badarulzaman, N., & Dawood, S. R. S. (2015). The Promise and Perils of the Island City of George Town (Penang) as a Creative City. *Urban Island Studies, 1*(1), 20–34.

Landry, C. (2008). *The Creative City: A Toolkit for Urban Innovators* (2nd ed.). Comedia Earthscan.

McCann, E. J. (2007). Inequality and Politics in the Creative City-Region: Questions of Livability and State Strategy. *International Journal of Urban and Regional Research, 31*(1), 188–196.

OECD. (2020, September 7). *Culture Shock: COVID-19 and the Cultural and Creative Sectors. OECD Policy Responses to Coronavirus (COVID-19)*. Retrieved from: http://www.oecd.org/coronavirus/policy-responses/culture-shock-covid-19-and-the-cultural-and-creative-sectors-08da9e0e/. Accessed 19 Nov 2020.

Pratt, A. (2011). The Cultural Contradictions of the Creative City. *City, Culture and Society, 2*, 123–130.

Sasaki, M. (2010). Urban Regeneration Through Cultural Creativity and Social Inclusion: Rethinking Creative City Theory Through a Japanese Case Study. *Cities, 27*(2010), S3–S9. Elsevier Ltd.

Index[1]

[1] Note: Page numbers followed by 'n' refer to notes.

S. L. Khoo, N. S. F. Chang, *Creative City as an Urban Development Strategy*, https://doi.org/10.1007/978-981-16-1291-6

F

G

L

O

The manufacturer's authorised representative in the EU is Springer Nature Customer Service Centre GmbH, Europaplatz 3, 69115 Heidelberg, Germany. If you have any concerns regarding our products, please contact ProductSafety@springernature.com

Printed and bound by CPI Group (UK) Ltd, Croydon, CR0 4YY

30/09/2026

02225326-0002